AF577044
WITHDRAWN

James W. Patterson graduated from the University of Florida in Gainesville in 1970 with a PhD in Environmental Engineering. Subsequently, he joined the faculty of the Illinois Institute of Technology in Chicago, where he became Director of the Pritzker Environmental Studies Center and Chairman of the IIT Department of Environmental Engineering.

Dr. Patterson is an active member of a number of professional societies in his field and acts in an advisory capacity to the Illinois Water Resources Center.

He also consults with government agencies and private engineering and industrial firms. During his career, he has authored or co-authored more than 50 publications and organized several workshops and symposia on air pollution, water pollution, energy and environmental quality.

WASTEWATER TREATMENT TECHNOLOGY

JAMES W. PATTERSON
Chairman
Department of Environmental Engineering
Illinois Institute of Technology
Chicago, Illinois

TD745
P32
1975

Second Printing, 1977

Copyright © 1975 by Ann Arbor Science Publishers, Inc.
P.O. Box 1425, Ann Arbor, Michigan 48106

Library of Congress Catalog Card Number 74-28653
ISBN 0-250-40086-3

Manufactured in the United States of America
All Rights Reserved

ROBERTS MEMORIAL LIBRARY
MIDDLE GEORGIA COLLEGE

73305

PREFACE

This volume developed out of work done over the period 1970–73 on behalf of the State of Illinois, to define the capabilities and limits of full-scale wastewater treatment technologies for 22 major industrial pollutants. The study culminated in a report to the State of Illinois in 1971, titled "Wastewater Treatment Technology." Based in part on the capabilities of technologies identified in that report, the Illinois Pollution Control Board established in 1972 effluent standards for industrial pollutants of concern. In order to confirm the data collected in the first report, a second study was performed in 1973 to update the original report, and incorporate new information. This revised second edition of the report "Wastewater Treatment Technology," was published by the Illinois Institute of Environmental Quality in 1973.

The demand for these reports was not anticipated, and as a result insufficient copies were printed to satisfy the requests. In light of the limited availability of the report, the rapid advancement of treatment technology in the field of industrial pollution control, and the continuing use and reference to the 1973 report by those who have obtained copies, a decision was made in 1974 to again update the study to reflect current technology, and to publish this volume. This, briefly, is the history leading to this text, *Wastewater Treatment Technology.*

The text contains 22 chapters, with each chapter dealing with a single pollutant. Each chapter should, ideally, accomplish three goals.

1. Identify sources and typical levels of the pollutant discussed.
2. Describe available treatment technology with regard to:
   a) how each technology operates and its limitations,
   b) what levels of treatment have been accomplished in full-scale treatment systems (preferably), or in pilot-scale systems if no full-scale data are available, and
   c) the economics of each technique, with regard to comparison of both capital investment requirement, and operating and maintenance costs.
3. Summarize concisely the major types of technology and effluent levels of pollutant achievable.

Where insufficient information was available to accomplish these goals, as for example for boron and selenium, such data as were available have been presented. An attempt has been made to fully reference all technical information contained herein, in order that the reader who wishes to delve more deeply into the original source literature may identify those sources and reports of particular interest.

Any credit for this text belongs not to this author, but to the many pollution control engineers and scientists who have assembled and published the results of their studies, observations, and experiences. Particular appreciation is due Dr. Roger A. Minear of The University of Tennessee, who was a vital member of the original study team on wastewater treatment technology. As any author soon realizes, a work such as this would be impossible without the cheerful forbearance and encouragement of his family, and for that I owe thanks to my wife, Ann, and my children Michael and Leslie.

This work required the evaluation and interpretation of the results of many workers, and for any errors in interpretation, or omissions, I bear full responsibility.

James W. Patterson
October, 1975

CONTENTS

1

TREATMENT TECHNOLOGY FOR ARSENIC

The principal aqueous forms of arsenic are the arsenite ion (AsO_2^-) and the arsenate ion (AsO_4^{-3}). In the presence of dissolved oxygen, arsenite is ultimately oxidized to arsenate (1). The oxidation state of the anion appears to have a significant effect upon treatment efficiency for arsenic removal.

Arsenic and arsenical compounds have been reported in wastewaters of the metallurgical industry, glassware and ceramic production, tannery operations, dye, and pesticide manufacture (2). Other industrial sources include the organic and inorganic chemicals and petroleum refining industries (3), and the rare-earth industry (4). Arsenic is widely associated with the manufacture of herbicides and pesticides (5). The manufacture of Paris green and calcium meta-arsenate, both insecticides, in one instance produced a wastewater containing 362 mg/l of arsenious oxide, As_2O_3 (6). In recent years, with the development of organic arsenicals, many of the inorganic arsenic-based forms have declined in use.

In addition to its use in pesticides, other industrial applications of arsenic include wood and hide preservatives, sheep dip, decolorizers in glass manufacturing, formulation of pigments, manufacture of lead shot, phosphate detergent builders and presoaks, and as a constituent of calcined phosphate used in many fertilizers. Hydrofluosilicic acid, commonly used as an agent in providing fluoride in water supplies, is reported to contain arsenic at 0.0097% (7).

Arsenic wastes from nonferrous smelting have been reported (3,8,9), although the main form appears to be as extremely small (1–3 μ diameter) particles from which arsenic can be profitably recovered (8). The soluble concentration of arsenic is dependent upon both chemical form and wastewater pH. For example a gold ore extraction process waste stream at pH 9.5 contained 900 mg/l particulate arsenic and only 10.1 mg/l of soluble arsenic (10). The soluble fraction, representing only 1% of the total arsenic present, was in the arsenite form. For a second waste stream at pH 3.1, particulate arsenic was 880 and soluble arsenic 132 mg/l (13%). Acid mine drainage of metal mines in the Rocky Mountains have been found to contain 6.0–22.0 mg/l soluble arsenic (11).

In one instance, wastewater from a sulfuric acid production plant was reported to contain 200–500 mg/l total arsenic (12). Sulfur dioxide used in the plant was obtained from a sulfide ore roasting process, and the high arsenic level presumably resulted from arsenic constituents of the ore.

High arsenic levels have been encountered in raw municipal water supplies, necessitating arsenic removal by the water treatment industry. Many ground water supplies in Central and South America contain excessive arsenic (13,14). Deep well waters in Taiwan, China, which contain up to 2.0 mg/l arsenic, are believed to be responsible for an endemic illness in the area called "black-foot" disease (15,16). Deep well waters in many desert areas of the southwestern United States contain arsenic at levels exceeding 0.1 mg/l (17).

TREATMENT TECHNOLOGY

Limited information is available on levels of arsenic in industrial wastes, and on current wastewater treatment processes and removals obtained. Much of the literature describing industrial sources and treatment of arsenic wastes is more than thirty years old. More up-to-date information is available on the removal of arsenic from drinking water, and in fact the methods for arsenic treatment of both drinking water and industrial wastes are similar. The treatment methods and arsenic removal efficiences discussed in detail below are summarized in Table 1.

Table 1. Summary of Arsenic Treatment Methods and Removals Achieved.

Treatment	Initial Arsenic (mg/l)	Final Arsenic (mg/l)	Percent Removal	Reference
Charcoal Filtration	0.2	0.06	70	18
Lime Softening	0.2	0.03	85	18
Lime Softening	0.5	0.03	95	10
Precipitation with Lime plus Iron	—	0.05	—	4
Precipitation with Alum	0.35	0.003–0.005	85–92	19
Precipitation with Ferric Sulfate	0.31–0.35	0.003–0.006	98–99	19
Precipitation with Ferric Sulfate	25.0	5	80	20
Precipitation with Ferric Chloride	3.0	0.05	98	21
Precipitation with Ferric Chloride	0.58–0.90	0.0–0.13	81–100	16
Precipitation with Ferric Hydroxide	—	0.6	—	22
Precipitation with Ferric Hydroxide	362.0	15–20	94–96	6
Ferric Sulfide Filter Bed	0.8	0.05	94	15
Precipitation with Sulfide	—	0.05	—	23
Precipitation with Sulfide	132.0	26.4	80	10

Major treatment methods for arsenic include sulfide precipitation (*e.g.*, as the sulfide or ferrisulfide), or complexation with polyvalent heavy metals such as ferric iron and coprecipitation with the metal hydroxide, plus adsorption into coagulant floc, with enmeshment of particulate arsenic. This second process is typical of the traditional coagulation process used in the water treatment industry. Other processes used, with varying degrees of success, include adsorption onto activated carbon and alumina, and ion exchange.

It has been reported that effluent arsenic levels of 0.05 mg/l are obtainable by precipitation of arsenic as the sulfide, with addition of sodium or hydrogen sulfide at pH 6–7 recommended (23). Sulfide precipitation has been found partially effective for arsenate, but ineffective for arsenite, in gold ore extraction plant effluent (10). At an optimal treatment pH of 7 and optional sulfide:arsenic ratio of 0.5, arsenate was only reduced from 132 to 26.4 mg/l. No precipitate resulted from sulfide treatment of arsenite wastewater.

Table 2 presents results from sulfide as well as other precipitating chemicals investigated (10). The results indicate that lime gave

Table 2. Precipitation Treatment of Arsenite and Arsenate in Gold Ore Extraction Wastewaters (10).

Precipitation Chemical	Arsenic Form	Optimal pH	Percent Removal	Optimal Ratio, Precipitation Ion: Arsenic Ion
Ferrous Sulfate	AsO_4^{-3}	8	94	$\frac{Fe^{+2}}{As} = 1.5$
	AsO_2^-	No Precipitation	0	
Ferric Chloride	AsO_4^{-3}	9	90	$\frac{Fe^{+3}}{As} = 4.0$
	AsO_2^-	8	95	
Alum	AsO_4^{-3}	7–8	90	$\frac{Al^{+3}}{As} = 4.0$
	AsO_2^-	7–8	95	
Sodium Sulfide	AsO_4^{-3}	7	80	$\frac{S^{=}}{As} = 0.5$
	AsO_2^-	No Precipitation	0	
Lime	AsO_4^{-3}	12	95	$\frac{Ca^{+2}}{As} = 9.8$
	AsO_2^-	12	95	
Caustic	AsO_4^{-3}	10	80	$\frac{Na^{+2}}{As} = 3.8$
	AsO_2^-	No Precipitation	0	

equal or better treatment than other chemicals, and at lower cost. It was effective on both arsenite and arsenate waste streams, since arsenite converts to the pentavalent form at high pH (10).

As indicated in Table 2, optimum dosage of ferric iron for arsenic removal was four-fold the wastewater arsenic concentration (10). Experience with a concentrated arsenic waste from a sulfuric acid plant indicated improved treatment at Fe^{+3}:As ratios up to 8, but no improvement at greater iron dosages (12). Increased coagulant dosage will result in greater sludge yield for ultimate disposal, however.

Although standard municipal sanitary sewage treatment processes do not remove arsenic, two common water treatment methods are effective in arsenic treatment. Use of the lime water-softening process was found to reduce an initial arsenic level of 0.2 mg/l down to approximately 0.03 mg/l, yielding 85% removal (18). Simple filtration through a charcoal bed yielded an effluent containing 0.06 mg/l of arsenic. This represented 70% removal of the original 0.2 mg/l. Rosehard and Lee reported 40% reduction of arsenite by activated carbon, from an initial concentration of 0.5–0.3 mg/l (10). This and other evidence (19,24) indicates that carbon is not as effective as other means of treating arsenic.

Arsenic removal has been observed in a conventional water treatment plant employing coagulation, settling, and filtration (20).

Ferric sulfate was employed as a coagulant, and at a coagulant dosage of approximately 500 mg/l it was possible to reduce the arsenic content of the water from 25 mg/l to 5 mg/l or less. The basis of this arsenic removal method is likely arsenic complexation with the iron, and simultaneous removal when the iron precipitates (5). This mechanism of coprecipitation is the probable explanation for soluble arsenic removals reported by all coagulation processes. Stable complexes of arsenate and arsenite form with most polyvalent metals, particularly iron, aluminum and zinc, and these complexes are captured by the hydroxide precipitation of these metals. Thus, when polyvalent metallic ions are present in or introduced into a wastewater, they complex with the arsenic ions and are coprecipitated at the pH of metal hydroxide formation. Treatment efficiency depends upon the types, concentrations and solubilities of polyvalent metal ions present, and the pH required to form hydroxide precipitates. Adsorption of arsenate or arsenite into the hydroxide floc may also occur, but likely plays a lesser role than coprecipitation. Chemical oxidation of arsenate to arsenite, prior to coagulation treatment, improves arsenic removal indicating that arsenate forms the more stable complex with the precipitating metal.

Ferrous hydroxide precipitation has also been employed to effectively remove arsenic from water supplies (14). The treatment sequence involved addition of ferrous sulfate for precipitation, and subsequent filtration. Effluent arsenic levels less than 0.05 mg/l was the goal of the treatment process. Ferrous coagulant treatment of arsenic-contaminated water has also been reported by others (13,22).

Logsdon, *et al.* (19) studied arsenate removal by precipitation with ferric sulfate and alum, in pilot plant tests. For initial arsenic concentrations of 0.31–0.35 mg/l, ferric sulfate treatment yielded 91–94% removal. Dual media filtration achieved an additional 5–7% removal for overall removals of 98–99% and effluent levels of 0.003–0.006 mg/l arsenic. Alum treatment was less effective, yielding 75–79% removal without filtration and 85–92% removal with filtration (19). Arsenite removals of only 10–25% were achieved by alum and 40–60% by ferric chloride. However, when chlorination preceded coagulant addition, removals equivalent to those reported for arsenate were observed (19). Chlorine apparently oxidized the arsenite to arsenate. For ferric sulfate precipitation, treatment efficiency remained constant from pH 5.5–8.5. Above this pH, efficiency declined. With alum optional precipitation pH was at 7.0, with slight reduction in efficiency at pH below 7.0, and rapid drop-off in removal efficiency at pH above 7.0. Similar pH effects for arsenate treatment by iron and aluminum coagulant salts have been observed in another study (1).

Experience with full-scale treatment of deep well water to remove arsenic indicates that both potassium permanganate and chlorine are effective oxidants of arsenite to arsenate (16). Copper sulfate coagulant performed poorly, as did alum, in precipitating the arsenic, while ferric sulfate and ferric chloride provided good removals. During four months of operation, arsenic levels of 0.36–0.56 mg/l were reduced to below 0.01 mg/l by chlorination to a chlorine residual of 0.1–2.2 mg/l, and ferric chloride dosages of 30–60 mg/l (16).

Arsenic in an industrial waste, which also contains heavy metals in solution, can be concurrently removed with precipitation of the heavy metals. Approximately 90% reduction of arsenic is to be expected (5). A patented waste recycle process removes arsenic and cadmium simultaneous with the precipitation of iron as ferric hydroxide (25).

In arsenic removal from wastewaters of the rare-earth industry, arsenic has been effectively removed by lime addition, but only in the presence of iron (4). Treatment levels below 0.05 mg/l were obtained. Similar experience was described by Nilsson (26), who was unable to achieve effluent arsenic levels below 9.7 mg/l through lime treatment of a low iron water.

Extensive pilot plant studies on physical-chemical removal processes for arsenic in municipal waste involved addition of a coagulant, followed by flocculation and settling, dual-media filtration, and carbon adsorption (27). Types and levels of coagulants employed are given in Table 3. Treatment results for an initial arsenic concentration of 5 mg/l are reported in Table 4. For this wastewater, filtration provided little or no improvement, although carbon column polishing was somewhat effective in all cases.

Both weak- and strong-base ion exchange resins appear effective in removing arsenate and arsenite from wastewater. Calmon (28), treating an arsenate waste containing 68 mg/l arsenic at pH 6.95, with a weak-base anion exchange resin (Ionic A-260), reported 82–100% removal. Medium- and strong-base resins (Ionac A-300, A-540, A-550) were less effective. Lee and Rosehart (10) reported successful treatment of arsenite wastes below pH 3 with Amberlite IR-45 (a weak base) and at pH 4–13 with Amberlite IRA 400 (a strong-base resin). Tests with an intermediate-base resin (Dowex Fps 4015L) showed optimum removal at pH 4.5–6, but overall removal was less than either the strong- or weak-base Amberlite resins (10).

Table 3. Treatment Systems for Arsenic Removal (27).

System	Coagulant	pH
Iron	Ferric sulfate @ 45 mg/l Fe	6.0
Low Lime	Lime @ 260 mg/l	10.0
	Ferric sulfate @ 20 mg/l Fe	
High Lime	Lime @ 600 mg/l	11.5

Table 4. Pilot Plant Arsenic Removal (27).

	Cumulative Percent Removal			Effluent Concentration
System	Settling +	Filtration +	Carbon	(mg/l)
Iron	90	89	96–98	0.06
Low Lime	79	79	82–84	0.92
High Lime	73	75	84–88	0.77

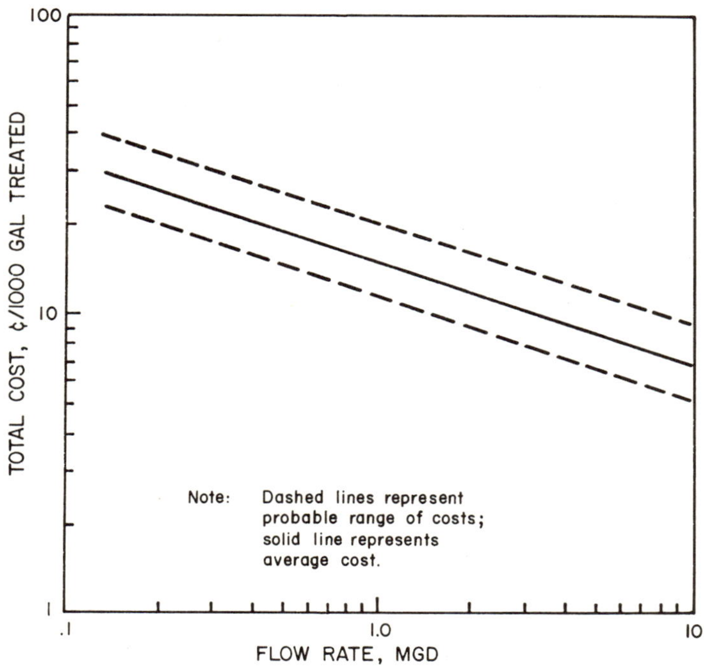

Figure 1. Total cost of treatment by coagulation, sedimentation and rapid sand filtration. Cost includes capital investment for 30 years at 4%, labor, power, chemicals, maintenance and repair, and heating of buildings (29).

The most commonly reported arsenic treatment process is one quite similar to that employed in treating drinking water. This is coagulant addition, sedimentation, and filtration. If the arsenic-bearing waste is acidic, pH adjustment to an alkaline condition may be required, and additional chemical costs are associated with that. Adjustment of pH is most frequently by use of lime, also employed in water treatment (specifically for water softening).

Therefore, it would appear that costs associated with routine water treatment processes would provide some basis for estimating the costs of arsenic removal. Figure 1 provides total treatment costs for a treatment process involving addition of coagulant, settling, and sand filtration. Costs of $0.15–0.40/1000 gal are indicated for plants treating 0.1–0.5 MGD. Figures 2 and 3 provide incremental lime costs for pH adjustment. Suggested guidelines for the selection of coagulants, based upon chemical cost and ease of sludge dewatering, are presented in Table 5.

SUMMARY

Arsenic is used as a raw material and produced as a waste in a variety of industrial processes. Despite this wide usage, there is

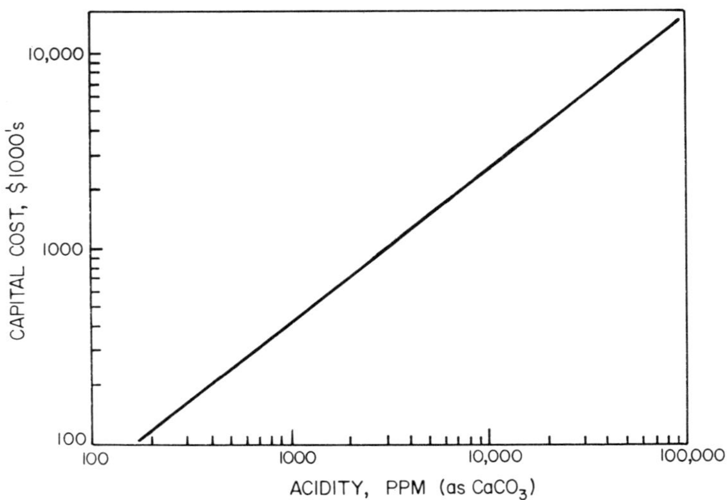

Figure 2. Capital cost for a 1-MGD lime neutralization plant, including sludge disposal (30).

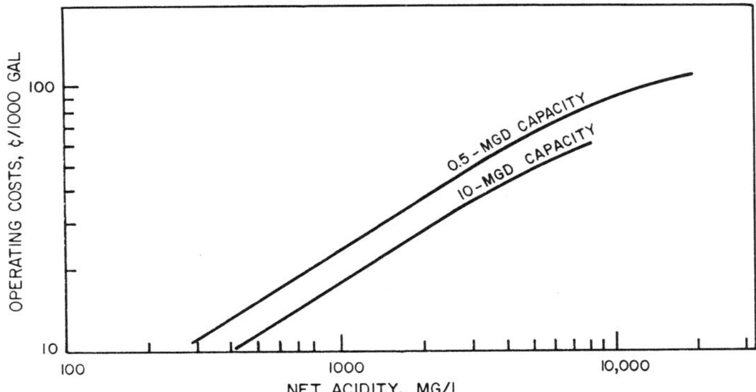

Figure 3. Operating costs for lime neutralization, including sludge dewatering by vacuum filtration (30).

Table 5. Chemical Coagulant Selection (31).

Chemical Process	Chemical Cost ($/dry lb)	Dosage Range (mg/l)	Resulting pH	Sludge Dewatering
High Lime	0.01	150–600	11.5–12.0	easy
Low Lime	0.01	75–250	9.5–10.5	easy
Very Low Lime	0.01	50–100	8.0–8.5	fair
Solid Alum	0.025	75–250	4.5–7.0	difficult
Liquid Alum	0.022	75–250	4.5–7.0	difficult
Ferric Chloride	0.045	35–150	4.0–7.0	fair
Ferrous Sulfate	0.019	70–200	4.0–7.0	fair

little information available, particularly in the current literature, on industrial waste levels and on industrial waste treatment processes. Removal of arsenic from municipal waters has been more widely reported. Arsenic treatment processes for both municipal waters and industrial wastes, insofar as the available literature indicates, are similar and commonly involve coprecipitation by addition of a polyvalent coagulant, to produce a hydroxide floc. If the arsenic is present as arsenite, effective treatment requires oxidation to arsenate by chlorine or permanganate addition, prior to coprecipitation. Based upon limited experience, sulfide precipitation, and both weak and strong base anion exchange are also effective treatment techniques. For low initial arsenic levels, these processes yield effluent arsenic levels of 0.05 mg/l or less. The treatment processes employed are standard operations, and values available in the literature for these processes (as applied to water and waste treatment) indicate treatment costs of $0.15–0.40/1000 gal.

REFERENCES

1. O'Connor, J. T. "Removal of Trace Inorganic Constituents by Conventional Water Treatment Processes," In *Proc., 16th Water Qual. Conf.*, **16**, 99–110, University of Illinois, Urbana, Illinois, February 13–14, 1974.
2. McKee, J. E. and H. W. Wolf. *Water Quality Criteria*, 2nd edition, California State Water Quality Control Board Publication No. 3-A, 1963.
3. Dean, J. G., F. L. Bosqui and K. H. Lanouette. "Removing Heavy Metals from Waste Water," *Environ. Sci. Technol.* **6**, 518–522, (1972).
4. Skripach, T., V. Kagan, M. Ramanov, L. Kamer and A. Semina. "Removal of Fluorine and Arsenic from the Wastewater of the Rare-Earth Industry," in *Proc. 5th Internat. Conf. Water Poll. Res.*, **2**, III-34, (New York: Pergamon Press, 1971).
5. Lund, H. F. *Industrial Pollution Control Handbook*, (New York: McGraw-Hill Book Co., 1971).
6. Cherkinski, S. N. and F. I. Ginzburg. "Purification of Arsenious Waste Waters," *Water Poll. Abstr.*, 14:315-316, 1941.
7. Newell, I. L. "Mercury and Other Heavy Metals in Water Supplies," *J. New Engl. Water Works Assoc.*, 85:3:289-295.
8. Swain, R. E. "Waste Problems in the Nonferrous Smelting Industry," *Ind. Eng. Chem.*, 31:1358-1361, 1939.
9. Stooff, H. and L. W. Haase. "Occurrence and Removal of Arsenic in Drinking Waters," *Chem. Abstr.*, 32:6370 (4), 1938.
10. Rosehart, R. and J. Lee. "Effective Methods of Arsenic Removal from Gold Mine Wastes," *Can. Mining J.*, June: 53-57, 1972.
11. Larsen, H. P., J. K. Shou and L. W. Ross. "Chemical Treatment of Metal-Bearing Mine Drainage," *J. Water Poll. Cont. Fed.*, 45:8: 1682-1695, 1973.
12. Ohsako, N. and H. Yoneyama. "Treatment of Arsenic-Containing Sulfuric Acid Production Plants," *Chem. Abstr.*, 78:235, 1973.
13. Trelles, R. A. and F. D. Amato. "Treatment of Arsenical Waters with Lime," *Water Poll. Abstr.*, 23:125, 1950.
14. Viniegra, G. and R. E. Marquez. "Chronic Arsenic Poisoning in the Lake Region: Section 4. Treatment of Drinking Water," *Water Poll. Abstr.*, 38:430-431, 1965.
15. Shen, Y. S. and C. S. Chen. "Relation Between Black-foot Disease and the Pollution of Drinking Water by Arsenic in Taiwan." in *Proc. 2nd Internat. Conf. Water Poll. Res., Tokyo*, 1:173-190, (New York: Pergamon Press, 1964).

16. Shen, Y. S. "Study of Arsenic Removal from Drinking Water," *J. Amer. Water Works Assoc.,* 651:543-548, 1973.
17. Bellack, E. "Arsenic Removal from Potable Water," *J. Amer. Water Works Assoc.,* 63:454-458, 1972.
18. Magnusen, L. M., T. C. Waugh, O. K. Galle and J. Bredfeldt. "Arsenic in Detergents: Possible Danger and Pollution Hazard," *Science* 168:389-390, 1970.
19. Logsdon, G. S., T. J. Sorg and J. M. Symons. "Removal of Heavy Metals by Conventional Treatment," in *Proc., 16th Water Qual. Conf.,* 16:111-133, University of Illinois, Urbana, Illinois, February 13-14, 1974.
20. Buswell, A. M., R. C. Gore, H. E. Hudson, H. C. Wiese and T. E. Larson. "Water Problems in Analysis and Treatment," *J. Amer. Water Works Assoc.,* 35:1303-1311, 1943.
21. Irukayama, K. "Discussion-Relation between Black-foot Disease and the Pollution of Drinking Water by Arsenic in Taiwan," in *Proc. 2nd Internat. Conf. Water Poll. Res., Tokyo,* 1:185-187, (New York: Pergamon Press, 1964).
22. Berezman, R. I. "Removal of Inorganic Arsenic from Drinking Water under Field Conditions," *Water Poll. Abstr.,* 29:185, 1965.
23. Curry, N. A. "Philosophy and Methodology of Metallic Waste Treatment," presented at 27th Purdue Industrial Waste Conference, Purdue University, 1972.
24. Lee, J. Y. and R. G. Rosehart. "Arsenic Removal by Sorption Processes from Waste Waters," *Can. Mining Metallurgical (CIM) Bull.,* 65:11:33-37, 1972.
25. Electrolytic Metal Corp. "Recovery of Manganese from Solutions Containing Iron, Manganese, Nickel, and Cobalt—Patent," *Chem. Abstr.,* 54:12960 (c), 1960.
26. Nilsson, R. "Removal of Metals by Chemical Treatment of Municipal Waste Water," *Water Res.,* 5:51-60, 1971.
27. Maruyama, T., S. A. Hannah and J. M. Cohen. "Removal of Heavy Metals by Physical and Chemical Treatment Processes," presented at 45th Ann. Conf., *Water Poll. Cont. Fed.,* Atlanta, Georgia, 1972.
28. Calmon, C. "Comment," *J. Amer. Water Works Assoc.,* 651:568-569, 1973.
29. Dorr-Oliver, Inc. "Cost of Waste Water Treatment Processes," Robert A. Taft Water Research Center Report, No. TWRC-6, U.S. Dept. Interior, Washington, 1968.
30. *The Economics of Clean Water, Vol. III, Inorganic Chemicals Industry Profile,* U.S. Dept. Interior, Washington, 1970.
31. Tofflemire, T. J. and L. J. Hetling. "Chemicals and Clarifiers: Which are Best?" *Water Wastes Eng.,* 10:11:F24-26, 1973.

2

TREATMENT TECHNOLOGY FOR BARIUM

Barium as barite (barium sulfate) is used industrially as a white pigment for paint (1). Barium is also employed in metallurgy, glass, ceramics and dyes manufacture, and in vulcanizing of rubber (2). Barium has been reported present in explosives manufacturing wastewater (3), and in midwestern ground waters at concentrations up to 10 mg/l (4).

TREATMENT TECHNOLOGY

In treatment of barium contained in an explosives manufacturing wastewater, barium was removed by precipitation as barium sulfate, upon addition of sodium sulfate (3). Coagulation of suspended solids in the explosives wastewater (including fine nonsettleable barium sulfate particles) with ferric sulfate provided solids removal. The process of precipitation of barium sulfate by addition of sodium sulfate is used by barite producers, to recover barium from solution (1). Barium sulfate is relatively insoluble, having a solubility product of 1×10^{-10}, and a maximum theoretical solubility of approximately 1.4 mg/l as barium at stoichiometric concentrations of barium and sulfate (5). The solubility level of barium can be reduced by adding sulfate in excess.

Others have also recommended sulfate precipitation for barium treatment, based upon the relative solubility products of barium sulfate and barium hydroxide (6,7). Alternately, in waters of moderate to high carbonate alkalinity, barium carbonate precipitation may be preferable. The solubility of barium carbonate is 1.6 x 10^{-9} (7). However, for effective carbonate precipitation, pH adjustment to above 10.5 is necessary.

Barium removal has been investigated in 4-gpm pilot plant studies (8). Three continuous-flow pilot plants received municipal wastewater dosed with barium (at 5 mg/l) and other metals, and removals by each unit process were determined. The unit processes were coagulation followed by settling, mixed-media filtration, and activated carbon adsorption. Table 6 lists the coagulants and operating pH's for each of the three pilot plants. The ferric sulfate dosages for the Iron and Low Lime systems are equivalent to respective sulfate dosages of 92.2 and 36.5 mg/l. Thus, sulfate was present in excess of the barium. Excess sulfate will reduce the barium solubility, and thereby improve the effectiveness of the barium sulfate precipitation process.

The results of the pilot plant studies are presented in Table 7.

Table 6. Pilot Treatment Systems for Barium Removal (8).

System	Coagulant	pH
Iron	Ferric Sulfate @ 45 mg/l Fe	6.0
Low Lime	Lime @ 260 mg/l Ferric Sulfate @ 20 mg/l Fe	10.0
High Lime	Lime @ 600 mg/l	11.5

Table 7. Pilot Plant Barium Removal (8).

System	Cumulative Percent Removal			Effluent Barium Concentration mg/l
	Settling +	Filtration +	Carbon	
Iron	97	95	95	0.27
Low Lime	99	99	99	0.03
High Lime	80	78	81	0.94

Treatment levels of 95% or greater were found for the two processes using ferric sulfate, while the High Lime process at pH 11.5 was much less efficient. The data of Table 7 indicate that increasing sulfate dosage may not be the only controlling treatment parameter, since the Low Lime system that employed sulfate at 36.5 mg/l was slightly more effective than the Iron system (at sulfate of 92.2 mg/l). The major fraction of removal was achieved by settling. Carbon adsorption did not improve treatment efficiency. The inefficiency of carbon in removing barium has also been noted by others (4,9).

The lack of correlation found in the pilot study between sulfate dosage and barium removal appears to be explainable in terms of the kinetics of barium sulfate precipitate formation. In batch studies on barium sulfate precipitation, high sulfate dosages apparently resulted in supersaturated barium sulfate solutions, which came to equilibrium with the crystalline phase very slowly (4). Coagulation with aluminum sulfate and ferric sulfate at coagulant dosages up to 120 mg/l yielded barium removal of 20–40% for initial barium concentrations of 7.0–8.6 mg/l and treatment pH 7.5–8.0. However, initial coagulant dosages of 100 mg/l, with subsequent (up to 24 hours later) coagulant dosages of 20 mg/l, yielded overall removals of 70% (4). These results indicate a very slow approach to solids equilibrium at high sulfate levels for barium sulfate precipitation, requiring long wastewater detention times.

Improved treatment has been achieved by lime softening treatment, with initial barium levels of 7–8 mg/l (4). Precipitation of barium carbonate is the most probable mechanism. Treatment efficiency was pH dependent, increasing to an optimum of 98% removal at pH 10.5, and dropping off at higher pH. These results are comparable to those described in Tables 6 and 7. At the optimum pH of 10.5, essentially all bicarbonate has been converted to carbonate. The reduced efficiency at pH above 11 may be due to preferential formation of the more soluble barium hydroxide.

Another effective barium treatment process is ion exchange. Successful ion exchange treatment has been reported for both nuclear wastes (10,11), and ground water softening (4). Ayres (10) cited extremely high removals of 99.999% for radioactive barium—140 by ion exchange, while Pressman, *et al.*, (11) report 99% removal of nonradioactive barium in a U.S. Army mobile ion exchange unit. In a full-scale ion exchange ground water softening plant, barium reduction from 11.7–0.17 mg/l, a 98.5% reduction, has been observed (7). These results indicate that ion exchange may be a more effective barium treatment technique than either sulfate or carbonate precipitation. However, successful ion exchange treatment would likely require use of general nonspecific water softening type resins, with consequent removal of constituents not removed by the precipitation technique. Further, ion exchange may represent a 4- to 9-fold greater treatment cost than chemical precipitation plus filtration (12).

SUMMARY

Methods of barium removal are summarized in Table 8, and include precipitation as barium sulfate upon addition of ferric or sodium sulfate, precipitation as barium carbonate at pH 10–10.5 with lime used for pH adjustment, and ion exchange. Extremely high removals were reported for the latter process. Based upon relative economics, however, and the low solubility of barium sulfate and barium carbonate, the precipitation method of barium removal would probably be the method of choice. Precipitation treatment levels of 0.03–0.27 mg/l have been reported in one study (8), while treatment costs of $0.20/1000 gal were given in a separate report (3).

Table 8. Summary of Barium Treatment Methods.

Method	pH	Barium Concentration (mg/l)		Percent Removal	Reference
		Initial	Final		
Precipitation					
Sulfate	6.0	5.0	0.27	95	8
Sulfate	10.0	5.0	0.03	99	8
Sulfate	—	—	0.5	—	13
Sulfate	7.5–8.0	7.0–8.6	2.1–2.6	70	4
Carbonate	10.5	7.0–8.0	0.15	98	4
Hydroxide	11.5	5.0	0.94	81	8
Ion Exchange	—	11.7	0.17	98	7
Ion Exchange	—	—	—	99.9+	10
Ion Exchange	—	—	—	99	11

REFERENCES

1. *The Economics of Clean Water, Vol. III, Inorganic Chemicals Industry Profile,* (Washington, D.C.: U.S. Dept. Interior, 1970).
2. McKee, J. E. and H. W. Wolf. *Water Quality Criteria,* 2nd edition, California State Water Quality Board Publication No. 3-A (1963).
3. "Taming Explosives," In *Ind. Eng. Chem.* **46** (12), 20A (1954).
4. Logsdon, G. S., T. J. Sorg and J. M. Symons. "Removal of Heavy Metals by Conventional Treatment," *Proc. 16th Water Qual. Conf.,* University of Illinois, Urbana, February 12-13, 1974.
5. Sillen, L. G. and A. E. Martel. *Stability Constants of Metal-Ion*

Complexes, 2nd edition, Special Publ. No. 17 (London, England: Chemical Society of London, 1964).
6. Culp, G. L. and R. L. Culp. *New Concepts in Water Purification,* (New York: Van Nostrand Reinhold Co., 1974).
7. Maruyama, T., S. Hannah and J. Cohen. "Removal of Heavy Metals by Physical and Chemical Treatment Processes," presented at *45th Annual Conf., Water Poll. Cont. Fed.,* 1972.
8. Sigworth, E. A. and S. B. Smith. "Adsorption of Inorganic Compounds by Activated Carbon," *J. Amer. Water Works Assoc.,* **64,** 386-391 (1972).
9. Ayres, J. A. "Treatment of Radioactive Waste by Ion Exchange," *Ind. Eng. Chem.* **43,** 1526-1531 (1951).
10. Pressmen, M., D. C. Lindsten and R. P. Schmitt. "Removal of Nuclear Bomb Debris; Strontium-90 - Yttrium-90, and Cesium-137 - Barium-137 From Water with Corps of Engineers Mobile Water Treatment Equipment," *Water Poll. Abstr.* **36,** 42-43, (1963).
11. Middlebrooks, E. J., J. H. Reynolds, G. L. and N. B. Jones. "Modifying Wastewater Treatment Plants to Met New Standards," *Publ. Works* **103** (8), 80-85 (1972).
12. Curry, N. A. "Philosophy and Methodology of Metallic Waste Treatment," presented at *27th Ind. Waste Conf.,* Purdue University, 1972.

3

TREATMENT TECHNOLOGY FOR BORON

Despite its widespread industrial use (Table 9), information on industrial waste levels of boron and on treatment processes for removal of boron from industrial wastewaters is scarce. Boron levels up to 5.5 mg/l have been reported in raw domestic sewage (1). A strong correlation between anionic detergent and boron concentrations was observed, indicating that the major source of the boron in this wastewater was sodium perborate, used as bleach in household washing powders. Boron normally occurs as the borate anion in water and wastewater.

Fluoborate solutions are sometimes used for plating of cadmium, copper, lead, nickel, tin, zinc and a number of alloys (2). They are used only for special purposes such as deposition of lead on battery bolts and terminals, plating of tin-lead alloys on bearings, or in cases where high-speed deposition is required for continuous wire or strip plating. Fluoborate plating solution concentrations (as BF_4) of 125,000–265,000 mg/l are common (2). This is equivalent to boron concentrations of 15,800–33,500 mg/l. Boric acid is used in nickel plating (3), and would presumably be a constituent of rinse water.

TREATMENT TECHNOLOGY

There appear to be few techniques to remove boron from aqueous solutions. One (patented) process for industrial wastewaters involves evaporation, and recondensing the vapor (4). The distillation process does not appear particularly effective, as it was observed that upon evaporation of a waste containing 21,000–22,000 mg/l of boron, the recondensed vapor still contained 50–80 mg/l of boron. Upon passing this water through a 6-foot column containing ceramic Raschig contact rings, the condensed vapor boron content was reduced to 2–3 mg/l.

Table 9. Industrial Uses of Boron.

Nuclear Reactor Core Solution	Rocket Fuels (boron hydrides)
Detergent Manufacture	Wire Drawing
Weatherproofing Wood	Fertilizers (boron-deficient soils)
Fireproofing Fabrics	Disinfectants
Manufacture of Glassware and Porcelain	Weed Control
	Welding and Brazing
Production of Leather and Carpets	Cutting Fluids
Cosmetics Manufacture	Plating
Photographic Supplies	Metallurgy (impact-resistant steels)

Based upon field observations and laboratory studies, the failure of several common treatment processes in reducing boron content of wastewater has been reported (1). The conventional biological waste treatment process has no effect on boron levels in domestic sewage. Lime, alum and ferric sulfate coagulants up to respective dosages of 400, 100 and 100 mg/l, are also ineffective. In one study, the reverse osmosis process removed only 10% of the total boron in a wastewater. This compared with 90–95% removal of chloride ion under similar operating conditions (1). Recent evidence indicates that boron removal by cellulose acetate membrane reverse osmosis is most efficient at pH 5, with removals of 38–60% (5). In reverse osmosis treatment of a brackish ground water initially containing borate at 0.35 mg/l as boron, reverse osmosis achieved a boron level of 0.14 mg/l in the permeate and 0.4 mg/l in the concentrate. This represents 60% rejection of the borate by the cellulose acetate membrane (5). Fletcher (6) has described a commercial solvent extraction reagent (Polyol, Dow Chemical Co.) which is used for boron extraction from brines.

A recently developed ion-exchange resin, Amberlite IRA-943 (Rohm and Haas) is claimed to be highly selective for removal of boron as borate and boric acid (7). The presence of other inorganic salts does not interfere significantly with the boron specificity of the resin. Limited experience in boron removal from sea water and irrigation water indicates that resin performance is independent of pH and ionic strength (7). For influent boron of 10 mg/l, effluent levels down to 1 mg/l were achieved. Removal efficiency was also dependent upon flow rate through the resin column. Calmon has quoted ion exchange costs of $0.12–0.30/1000 gal for boron treatment (8).

SUMMARY

Reported treatment methods from boron include evaporation, reverse osmosis and ion exchange. Evaporation with vapor condensation and Raschig ring column contact produced an effluent boron level of 2–3 mg/l (4). Reverse osmosis at optimum process pH had a rejection efficiency of 60% and an effluent boron level of 0.14 mg/l (5). Ion exchange has achieved 90% removal and an effluent boron level of 1 mg/l (7).

REFERENCES

1. Waggott, A. "An Investigation of the Potential Problem of Increasing Boron Concentrations in Rivers and Water Courses," *Water Res.* **3**, 749-765 (1969).
2. Lowe, W. "The Origin and Characteristics of Toxic Wastes, with Particular Reference to the Metal Industries," *Water Poll. Cont. (London)* 270-280 (1970).
3. Rudolfs, W., Ed. *Industrial Waste Treatment,* (New York: Van Nostrand Reinhold Co., 1953).
4. Doeldner, R. W. "Vapor-Compression Distillation of Nuclear Reactor Coolant Containing Boron Compounds, U.S. Patent 3,480,515," *Chem. Abstr.* **72,** 2748m, (1970).
5. Cruver, J. E. "Reverse Osmosis—Where It Stands Today," *Water Sewage Works* October, 74-78 (1973).
6. Fletcher, A. W. "Metal Extraction from Waste Materials," *Chem. Ind.* **28,** 776-780 (1971).
7. Kunin, R. "A Macroreticular Boron-Specific Ion-Exchange Resin," In *Trace Elements in the Environment,* E. L. Kothny, Ed., Advances in Chemistry Series 123 (Washington, D.C.: American Chemical Society, 1973).
8. Calmon, C. "Trace Heavy Metal Removal by Ion Exchange," In *Traces of Heavy Metals in Water-Removal Processes and Monitoring,* U.S. Environmental Protection Agency, Report EPA-902/9-74-001 (Washington, D.C.: U.S. Government Printing Office, 1974).

4

TREATMENT TECHNOLOGY FOR CADMIUM

Potential sources of cadmium wastewaters include metallurgical alloying, ceramics manufacture, electroplating, inorganic pigments, textile printing, chemical industries, and mine drainage (1,2). Of the total industrial cadmium use, 90% is utilized in electroplating, pigments, plastics stabilizers, alloying and batteries. Most of the remaining 10% is used for television tube phosphors, golf course fungicides, rubber-curing agents and nuclear reactor shields and rods (3).

High concentrations of cadmium have been reported in plating wastes and mine drainage (Table 10). Cadmium discharge from inorganic pigment industry plants ranges from 2–120 g/day, with a plant average for plants using cadmium of 8 g/day (9). The principal pigments are cadmium sulfide and cadmium selenide.

TREATMENT TECHNOLOGY

Methods of cadmium wastewater treatment currently in use are essentially those employed for most heavy metal wastes encountered in the plating and metal processing industries. Common methods include precipitation as the hydroxide or sulfide, and ion exchange.

Table 10. Cadmium Concentrations Reported for Plating and Mine Drainage Wastewaters.

Process	Cadmium Concentration (mg/l)	Reference
Automobile Heating Control Manufacturing	14–22	4
Plating Rinse Waters		
Automatic barrel zinc & cadmium plant	10–15	5
Mixed manual barrel and rack	7–12	5
Plating Rinse Waters (large installations)	15 ave., 50 max.	6
Plating Rinse Waters		
0.5 gph dragout	48	7
2.5 gph dragout	240	7
Plating Bath	23,000	7
Bright Dip and Passivation Baths	2,000–5,000	5
Acid Lead Mine Drainage	1,000	1
Acid Mine Drainage	440–1,000	8

Table 11. Treatment Technology for Cadmium Removal.

Method	Treatment pH	Initial Cd (mg/l)	Final Cd (mg/l)	Percent Removal	Reference
Hydroxide Precipitation	8.0	—	1.0	—	10
Hydroxide Precipitation	10.0	—	0.10	—	10
Hydroxide Precipitation	11.0	—	0.00075	—	11
plus Filtration	11.0	—	0.00070	—	11
Hydroxide Precipitation	9.0	—	0.54	—	12
Hydroxide Precipitation plus Filtration	15.5	—	0.014	—	13
Hydroxide Precipitation plus Filtration	—	—	0.08	—	14
Coprecipitation with Ferrous Hydroxide	6.0	—	0.050	—	13
Coprecipitation with Ferrous Hydroxide	10.0	—	0.044	—	13
Sulfide Precipitation	6.5	440–1000	0.008	99+	8
Reverse Osmosis	11.5	—	—	78–99+	15
Freeze Concentration	—	100	0.4	99.6	16

Although not frequently mentioned in the cadmium literature, electrolytic recovery and evaporative recovery processes, as used for copper, silver and nickel waste recovery, are technically feasible, but only for a concentrated waste or after a preconcentration step such as ion exchange. Effluent values for cadmium treatment techniques are summarized in Table 11.

Chemical Precipitation

Cadmium forms an insoluble and highly stable hydroxide at alkaline pH (17). Most effective cadmium hydroxide precipitation occurs between pH 9.5 and 12.5. Freshly precipitated cadmium hydroxide has been reported to leave approximately 1 mg/l of residual cadmium ion in solution at pH 8, but this is reduced to 0.1 mg/l at pH 10 (10). Precipitation at pH above 11 has been reported to yield an effluent cadmium level of 0.00075 mg/l. Sand filtration slightly reduced this residual cadmium concentration to 0.00070 mg/l (11). It has been indicated that coprecipitation with iron hydroxide at pH 8.5 improves cadmium removal (18,19), and that coprecipitation with aluminum hydroxide is also effective for cadmium, as well as zinc and nickel (11).

Evidence for the beneficial treatment effect of coprecipitation with iron hydroxide is presented by Maruyama, *et al.* (13). In pilot plant studies of metals in municipal wastewater, three treatment techniques were investigated. These were ferrous sulfate (45 mg/l Fe) addition at pH 6, low lime (260 mg/l) plus ferrous sulfate (20 mg/l Fe) at pH 10 and high lime (600 mg/l) at pH 11.5. Flocculation and settling were followed by mixed media filtration prior to passage through activated carbon columns. With iron addition only, a residual cadmium concentration of 0.05 mg/l was reported. Low lime treatment yielded 0.044 mg/l and high lime 0.014 mg/l cadmium residuals. Thus, iron coprecipitation alone at pH 6 yielded near-equal results to iron plus lime treatment at pH 10.

In hydroxide precipitation of cadmium from a waste stream at a machine plating plant, an effluent pH of 9.0 and cadmium concentration of 0.54 mg/l was reported for the lime precipitation process (12). Although sand filtration was not used, the process incorporated coagulant aid addition at 1–2 mg/l. The effluent contained only 2 mg/l of suspended solids, indicating efficient solids removal by sedimentation. Extremely effective cadmium removal has been achieved in a pilot plant lime coagulation-settling process. Treating combined sewage effluent from a secondary treatment plant, the

73305

ROBERTS MEMORIAL LIBRARY
MIDDLE GEORGIA COLLEGE

process achieved 94.5% cadmium removal for the trace quantities present (20). A similar pilot plant study on municipal waste secondary effluent used lime, flocculation, settling and ammonia stripping followed by recarbonation, mixed media filtration and activated carbon (21). Flows ranged from 7000–35,000 gpd and influent cadmium levels of 0.011–0.130 mg/l were reduced to 0.000–0.005 mg/l.

In the presence of complexing agents (*e.g.*, cyanide) it is difficult to precipitate the cadmium ion (19). In the event that appreciable complexing agents are present in the waste solution, pretreatment to destroy the agents is required. Cadmium plating wastes normally contain cyanide. Effective precipitation of cadmium from spent plating baths or rinse waters is, therefore, dependent upon prior cyanide removal. Fortunately, cyanide breakdown is relatively rapid and easy, as has been demonstrated for zinc and copper cyanide baths (22, 23).

The availability of a cyanide-free cadmium plating bath with acceptable product quality has been suggested as an answer to simplified treatment needs (24). However, the bath is comprised of high ammonium salt concentration (112 g/l) and also contains a proprietary chelating agent. Problems related to cadmium complexation by both the ammonia and proprietary chelating agent could be anticipated from such a system (25).

In the absence of, or after destruction of, cyanide, unit operations involved in cadmium removal are similar to those of the conventional tertiary treatment sequence of coagulation, settling, and sand filtration. The latter step has become of increasing importance for complete removal of nonsettling and slow settling metal hydroxide flocs (11). Figure 4 provides approximate cost data for this sequence of operations. Simple precipitation treatment costs are approximately $0.25/1000 gal without filtration and $0.45/1000 gal with filtration (18). Since cadmium treatment is likely to be integrated with other treatment schemes, generalized cost data for precipitation systems would apply unless high treatment pH (*i.e.*, above 10) becomes necessary. Higher lime costs would be expected under such conditions.

The impact of effective solids removal on effluent cadmium concentrations has been demonstrated. Chalmers (14) reported treated effluent cadmium concentration of 0.7 mg/l after cyanide-chromium treatment plus neutralization and sand filtration of a 10,000 gph process. Cadmium was further reduced to 0.08 mg/l using 10-hour settling followed by paper press filtration (as opposed to sand filtration). The latter process could handle only 30% of the flow and was deemed very costly.

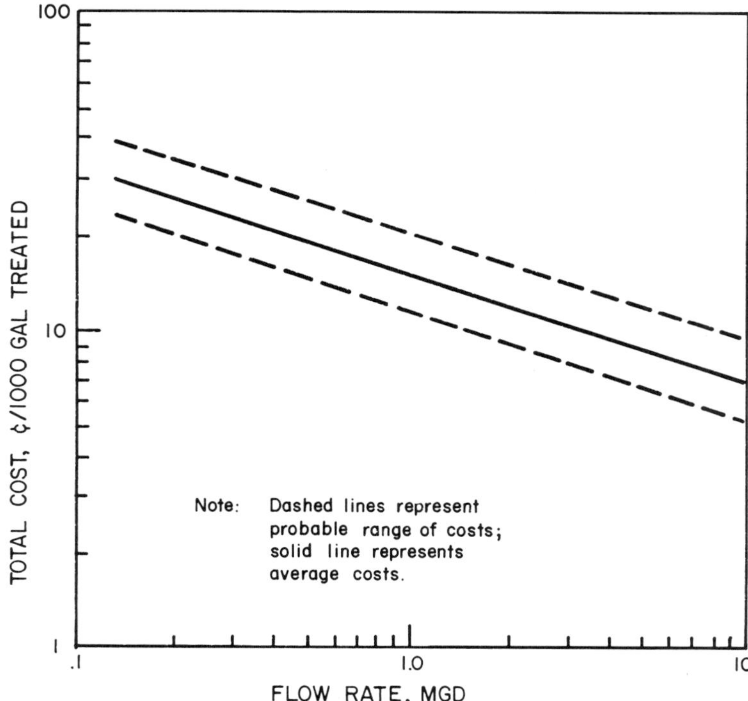

Figure 4. Total cost of treatment by coagulation, sedimentation, and rapid sand filtration. Cost includes capital investment for 30 years at 4%, labor, power, chemicals, maintenance and repair, and heating of buildings (26).

A proprietary hydrogen peroxide oxidation-precipitation system (the Kastone process) has been developed, which simultaneously oxidizes cyanides and precipitates the oxide of cadmium (rather than the hydroxide). Cadmium oxide precipitate is claimed to be much easier to remove from solution than cadmium hydroxide (27). This process is reported as suitable for small plating operations, since solids removal is readily accomplished by slight modification to simple filtration apparatus already present in most shops.

Hydroxide precipitation with lime may be unsatisfactory for cadmium, due to the high pH required for effective treatment (28). The use of sulfide for cadmium precipitation appears feasible, although it has had only limited application (8,29). In one application, acid mine drainage, initially at pH 2.6 and containing cadmium at 440–1000 mg/l, has been treated by two-stage precipitation. In the first stage, the wastewater was neutralized with lime to pH 5.0. In the second stage, pH was further adjusted to 6.5, and sodium sulfide added to precipitate cadmium and other heavy metals. Effluent cadmium of 0.008 mg/l was achieved (28).

Ion Exchange

There are many ion exchange resins with high specificity for cadmium (30). Ion exchange can be used as a polishing treatment or recovery process. The concentrated solution obtained upon regeneration of the ion exchange resin is often suitable for economical recovery procedures. The recovery value of cadmium is estimated at $1.20–$6.00/1000 gal, at a solution concentration of 50–250 mg/l of cadmium (26). The extra capital cost of an ion exchange treatment plant can be offset in one-half to two years when product recovery value is substantial, as is the case with cadmium (30). Ion exchange is unsuitable, however, for recovery of mixed cadmium cyanide solutions (31). Depending upon the ion exchange resin and regeneration configuration, installed costs may be $411–464/1000 gpd capacity, with operating costs of $0.52–$0.66/1000 gal treated (32).

Evaporative Recovery

A single effect 100-gph evaporator has been installed on a 5-barrel, 800-gallon bath cadmium plating line in one plant (33). Economics for the cadmium system were not available, but a similar zinc recovery system in the same plant had proven to be successful. Since the recovery value of cadmium plating solution is greater than zinc, similar results were to be expected. Culotta and Swanton (34) have compared capital and operating costs for various evaporator systems. Single-effect evaporators for a 150-gph unit represent a capital investment of $13,889/1000 gpd capacity (24-hr basis), and treatment costs are $17.40/1000 gal, based on 4000 hr/yr operation. Chemical recovery and water reuse values obviously must offset these costs, for evaporative recovery to become economically feasible.

The recovery process necessitates waste stream segregation if water and chemical reuse are to offset capital and operational costs. The recovery system is not restricted to evaporation, since any method can be used which will concentrate the waste solution to the extent that it can be reintroduced to the process chemical reservoir, while isolating water for rinse reuse. Reverse osmosis has been examined for treatment of cadmium plating rinse solution (15). Cadmium removal of 78–99+% was reported, with cyanide removals of 10–97%. Electrodialysis has been claimed to be a successful and economical means of treating wastewater in the plating industry but no data are available in support of the claim (35). Freeze recovery has also been considered and reported effective. A 2500-gpd pilot plant unit has achieved 99.6% reduction of cadmium, to a final concentration of 0.4 mg/l (16).

SUMMARY

Removal of cadmium from a waste stream can be accomplished by precipitation as the hydroxide at elevated pH, but values of pH 10 or greater are necessary for effective removal in the absence of coprecipitation with iron or aluminum hydroxide. Removal solely on a solubility basis is restricted to 0.1 mg/l residual cadmium at pH 10, although greater efficiency is possible at higher pH. Coprecipitation with, or adsorption on ferrous hydroxide floc may enhance removal. Sulfide precipitation, at more neutral pH, appears feasible. Organic or inorganic complexing agents in the waste stream will reduce the effectiveness of precipitative removal.

A scarcity of operating data and costs specifically related to cadmium treatment exists in the literature. However, costs associated with the various treatment procedures for other nonferrous heavy metals should apply to cadmium solutions of similar general characteristics due to the similarity of treatment methods. Recovery processes would be expected to be more attractive than with many nonprecious metal wastes, due to the high value of cadmium metal.

REFERENCES

1. McKee, J. E. and H. W. Wolf. *Water Quality Criteria,* 2nd edition, California State Water Quality Control Board, Publication No. 3-A, 1963.
2. Santaniello, R. M. “Air and Water Pollution Quality Standards, Part 2: Water Quality Criteria and Standards for Industrial Effluents,” *Industrial Pollution Control Handbook,* Herbert F. Lund, editor, (New York: McGraw-Hill Book Co., 1971).
3. “Metals Focus Shifts to Cadmium,” In *Environ. Sci. Technol.,* **5,** 754-755 (1971).
4. Gard, S. M., C. A. Snavely and D. J. Lemon. “Design and Operation of a Metal Wastes Treatment Plant,” *Sew. Ind. Wastes* **23,** 1429-1438 (1951).

5. Lowe, W. "The Origin and Characteristics of Toxic Wastes with Particular Reference to the Metals Industries," *Water Poll. Cont. (London)* 270-280 (1970).
6. Pinkerton, H. L. "Waste Disposal. Inorganic Wastes," In *Electroplating Engineering Handbook,* 2nd edition, A. Kenneth Graham, Editor, (New York: Van Rostrand Reinhold Co., 1962).
7. Nemerow, N. L. *Theories and Practices of Industrial Waste Treatment* (Reading, Massachusetts: Addison-Wesley Publishing Co., 1963).
8. Larsen, H. P., J. K. P. Shou and L. W. Ross. "Chemical Treatment of Metal-Bearing Mine Drainage," *J. Water Poll. Cont. Fed.* **45** (8), 1682-1695 (1973).
9. Barrett, W. J., G. A. Morneau and J. J. Roden. "Waterborne Wastes of the Paint and Inorganic Pigments Industries," U.S. Environmental Protection Agency Report EPA-670/2-74-030 (Cincinnati, Ohio: U.S. Government Printing Office, 1974).
10. Jenkins, S. H., D. G. Knight and R. E. Humphreys. "The Solubility of Heavy Metal Hydroxides in Water, Sewage and Sewage Sludge, I. The Solubility of Some Metal Hydroxides," *Internat. J. Air Water Poll.* **8**, 537-556 (1964).
11. Culp, G. L. and R. L. Culp. *New Concepts in Water Purification,* (New York: Van Nostrand Reinhold Co., 1974).
12. Hansen, N. H. and W. Zabban. "Design and Operation Problems of a Continuous Automatic Plating Waste Treatment Plant at Data Processing Division, IBM, Rochester, Minnesota," *Proc. 14th Purdue Ind. Waste Conf.,* pp. 227-249, 1959.
13. Maruyama, T., S. A. Hannah and J. M. Cohen. "Removal of Heavy Metals by Physical and Chemical Treatment Processes," presented at 45th Annual Water Poll. Cont. Fed. Meeting, 1972.
14. Chalmers, R. K. "Pretreatment of Toxic Wastes," *Water Poll. Cont. (London)* 281-291 (1970).
15. Donnelly, R. G., R. L. Goldsmith, K. J. McNulty and M. Tan. "Reverse Osmosis Treatment of Electroplating Wastes," *Plating* **61** (5), 432-442 (1974).
16. Campbell, R. J. and D. K. Emmerman. "Recycling of Water from Metal Finishing Wastes by Freezing Process," presented at the Reuse Treatment of Waste Water Symposium, ASME/EPA/AICHE, New Orleans, Louisiana, March 28, 1972.
17. Cotton, F. A. and G. Wilkinson. *Advanced Inorganic Chemistry* (New York: Wiley-Interscience Publishers, 1962).
18. Anderson, J. S., J. M. Phillips and C. B. Schriver. "Water Contamination by Certain Heavy Metals," presented at 44th Annual Meeting, Water Poll. Cont. Fed., 1971.
19. Weiner, R. F. "Acute Problems in Effluent Treatment," *Plating* **54,** 1354-1356 (1967).

20. Linstedt, K. D., C. P. Houck and J. T. O'Connor. "Trace Element Removals in Advanced Wastewater Treatment Processes," *J. Water Poll. Cont. Fed.* **43**, 1507-1513 (1971).
21. Argo, D. G. and G. L. Culp. "Heavy Metals Removal in Wastewater Treatment Processes: Part 2—Pilot Plant Operation," *Water Sewage Works* **119**, 128-132 (1972).
22. Eden, G. E., B. L. Hampson and A. B. Wheatland. "Destruction of Cyanide in Waste Waters by Chlorination," *J. Soc. Chem. Ind.* **69** (8), 244-249 (1950).
23. Lancy, L. E. "An Economic Study of Metal Finishing Waste Treatment," *Plating* **54**, 157-161 (1967).
24. Beckwith, M. M. "Cyanide-Free Cadmium Plating Bath," *Plating/ Finishing Pract.* **58**, 413-415 (1971).
25. Minear, R. A. and J. W. Patterson. "Treatment of Metallic Wastewaters," In *Water Pollution in Metropolitan Areas,* Proceedings of a Symposium at Illinois Institute of Technology, November 30, 1972, pp. 110-139.
26. Digregorio, D. "Cost of Wastewater Treatment Processes," Robert A. Taft Water Research Center Report No. TWRC-6, (Washington, D.C.: U.S. Dept. Interior, 1968).
27. "New Process Detoxifies Cyanide Wastes," *Environ. Sci. Technol.* **5**, 496-497 (1971).
28. Dean, J. D., F. L. Bosqui and K. H. Lanouette. "Removing Heavy Metals from Waste Water," *Environ. Sci. Technol.* **6**, 518-522 (1971).
29. Shimoiizaka, J. "Recovery of Xanthate from Cadmium Xanthate," *Nippon Kogyo Kaishi (Japan)* **88**, 539-43 (1972).
30. Mattock, G. "Modern Trends in Effluent Control," *Metal Finishing J.* **14**, 168-175 (1968).
31. Oldan, K. and T. C. Hesler. "Profitable Recovery of Plating Wastes by Reconcentration of Reduced Volume Rinses," *Plating* **43**, 1022-1025 (1956).
32. Zievers, J. F. and C. J. Novotny. "Recovery of Mixed Rinse Water by Means of Ion Exchange," *Plating* **58**, 482-485 (1971).
33. Gallo, B. R. and J. M. Culotta. "Save on Plating Waste System," *Water Wastes Eng.* **9**, A18-A19 (1972).
34. Culotta, J. M. and W. F. Swanton. "Recovery of Plating Wastes: Selection of Lowest Cost Evaporator," *Plating* **57**, 1221-1223 (1970).
35. Bovet, E. D. "Industrial Waste Desalting for Byproduct Recovery," *J. Amer. Water Works Assoc.* **62**, 539-542 (1970).

5

TREATMENT TECHNOLOGY FOR CHLORIDE

Chlorides are present in practically all waters. They may be of natural origin, or derived (a) from sea water contamination of ground water supplies, (b) from salts spread on fields for agricultural purposes, (c) from human or animal sewage, or (d) from industrial effluents, such as those from food processing, paper works, galvanizing plants, water softening plants, oil wells, and petroleum refineries (1).

Waste chloride concentration from one Kraft process paper mill has been reported to range from 350–1760 mg/l (2). Average-sized steel mills using hydrochloric acid pickling discharge from their pickling line about 9700 lb/day of chloride as the ferrous salt, and 4500 lb/day as hydrochloric acid. The waste stream from the pickling line will contain a total chloride content of approximately 7000 mg/l (3).

Oil refineries and petrochemical processes are major producers of high chloride wastes. Crude oil desalting is a significant source of chloride in petroleum refinery wastes. One refinery effluent has been described as containing total dissolved solids (TDS) of 1163 mg/l, primarily as sodium chloride dissolved out of crude oil by desalting (4). Petrochemical processes, including chlorination, oxidation, polymerization and alkylation produce chloride wastes, usually associated with calcium or aluminum (5).

The Solvay process, widely used in making soda ash, yields wastewaters high in chloride. Soda ash is a major raw material of many industries, including glass, chemicals, pulp and paper, soap and detergent, aluminum and water treatment. The Solvay process uses sodium chloride brine solution, and yields a substantial amount of waste calcium chloride (6). The chloride content of a typical soda ash manufacturing waste is 90,000 mg/l (7). While some calcium chloride is recovered by distillation, this practice is not sufficiently prevalent to prevent the Solvay process having a major waste disposal problem (6).

Food processing industry wastewaters may also be high in chloride content. One 400,000 gpd waste flow from an olive processing facility in California contained 3500–6000 mg/l of chloride (8). The chloride represented 35–40% of the total dissolved solids concentration of that waste. Another report has cited chloride levels of 36–4280 mg/l of chloride from olive processing, and 24,220 mg/l of chloride from cucumber pickling (9).

Frequently, chloride ion is added to a wastewater as a result of waste treatment required to remove some less desirable constituent. Sodium chloride and hydrochloric acid are often used to regenerate spent ion exchange resins used in water purification. Chloride ions on the regenerated resins then exchange with undesirable anions as the wastewater passes through the ion exchange bed. Another example of chloride addition is chlorination treatment of cyanide waste streams.

TREATMENT TECHNOLOGY

Two of the most frequently employed methods of controlling high chloride wastes involve techniques that more closely achieve ultimate disposal than chloride reduction *per se*. These methods are deep well injection and holding basins, including solar evaporating ponds. Methods of actually removing chloride ion from a waste stream include anion exchange and electrodialysis, in addition to other broadly applicable, total dissolved solids reduction methods discussed in Chapter 21.

Deep Well Injection

Deep well disposal has been employed for a chemical industry wastewater containing 10–15% sodium chloride at volumes of 500–600 gpm. The waste stream also contained dissolved trace metals and trace organics (10). Other reports of deep well disposal of high chloride wastes are also in the literature (5). Figures 5 and 6 present capital investment and operating costs for deep well disposal. Injection pressures of 200–5000 psi have been reported for chloride wastes (5). In light of the hazards and uncertainties associated with subsurface waste disposal, deep well injection may create pollutional problems far more severe than the ones being corrected (11).

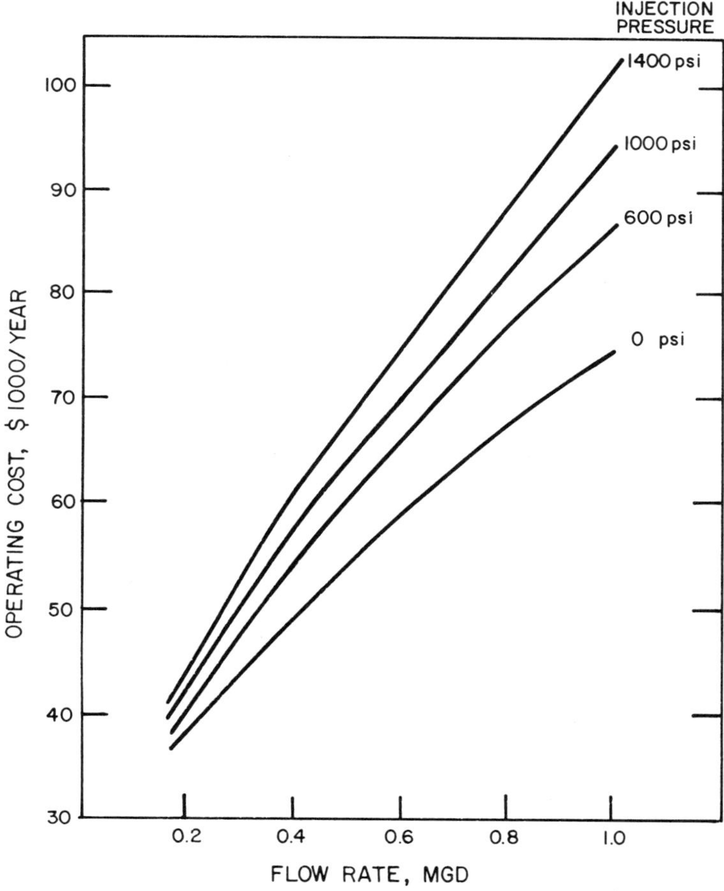

Figure 5. Annual operating costs for waste disposal by deep well injection systems (6).

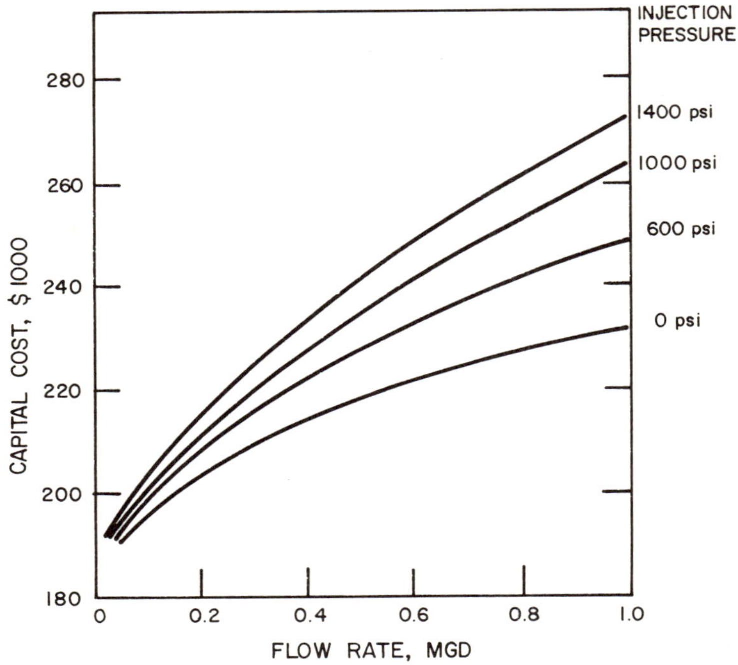

Figure 6. Capital costs for deep well injection waste disposal systems (6).

Ponds and Basins

The concepts behind the use of holding ponds and solar evaporating basins are similar. Holding ponds are used to store wastewater in times of low-receiving stream flow, with gradual wastewater release to the stream in high-flow periods at rates necessary to maintain acceptable stream quality standards. Solar evaporating ponds utilize the slow loss of water by evaporating from the sun's heat to avoid any release of wastewater at all the receiving streams. Disposal by evaporation is limited by geographical location, climate and land availability (12). It is obviously a very attractive method where stringent effluent regulations are in force, and geographical, climatic and land conditions are favorable. Pollutant containment is essentially complete providing floating oils, which retard evaporation, are absent. Table 12 presents capital and operating costs for evaporation ponds as employed in petroleum refinery waste treatment.

Table 12. Costs of Evaporative Pond Treatment of Wastewaters (12).

Waste Flow (MGD)	Capital Costs ($)	Annual M&O Costs ($/yr)
3.0	56,250	575
7.5	90,000	900
15.0	127,500	1275

Jones (7) has reported the use of a holding pond and controlled discharge of a chloride waste, where underground disposal was prohibited. The waste stream contained approximately 90,000 mg/l of chloride. A holding basin, having a capacity of 150,000 cu ft, was constructed at a cost of approximately $1,000,000 (including land, 1958 prices). Discharge of chloride waste from this basin was proportionate to receiving river flow. Evaporating ponds have been employed in California to retain olive processing brine wastes. The wastes contain 10,000–15,000 mg/l of dissolved solids, of which chloride constituted 35–40% (8).

Ion Exchange

Higgins (13) has discussed the use of ion exchange to remove chlorides, and points out that removal of strong acid and acidic ions such as chloride, fluoride and phosphate anions requires a weak-base type anion exchange resin. Strong-base anion exchange resins are more selective for common ions such as hydroxide and bicarbonate, than for chloride, and thus less effective for chloride treatment. Chloride removal is not complete; a recent report indicating only approximately 80% chloride reduction upon ion exchange treatment of wastes containing chloride at 600–3600 mg/l (9). Other disadvantages include limitations on wastewater dissolved solids, which should be less than 2000 mg/l, and a highly alkaline wastewater being required for good treatment (14).

When the resin is exhausted, exchange capacity can be regenerated with weak alkali-ammonia solution (13). Figure 7 presents amortization cost (based on 7-year write-off at 7%) versus plant size, and chemical (regenerating) costs versus wastewater TDS concentration for a weak base ion exchange plant. If total dissolved solids were due solely to the presence of sodium chloride, 100 mg/l TDS would be equivalent to 60.7 mg/l chloride, and likewise, 10 times as much TDS would equal a 10-fold increase in chlorides. Equivalent cost data have been reported elsewhere for ion exchange treatment (6). Capital cost, for a 100,000 gal/day anion exchange

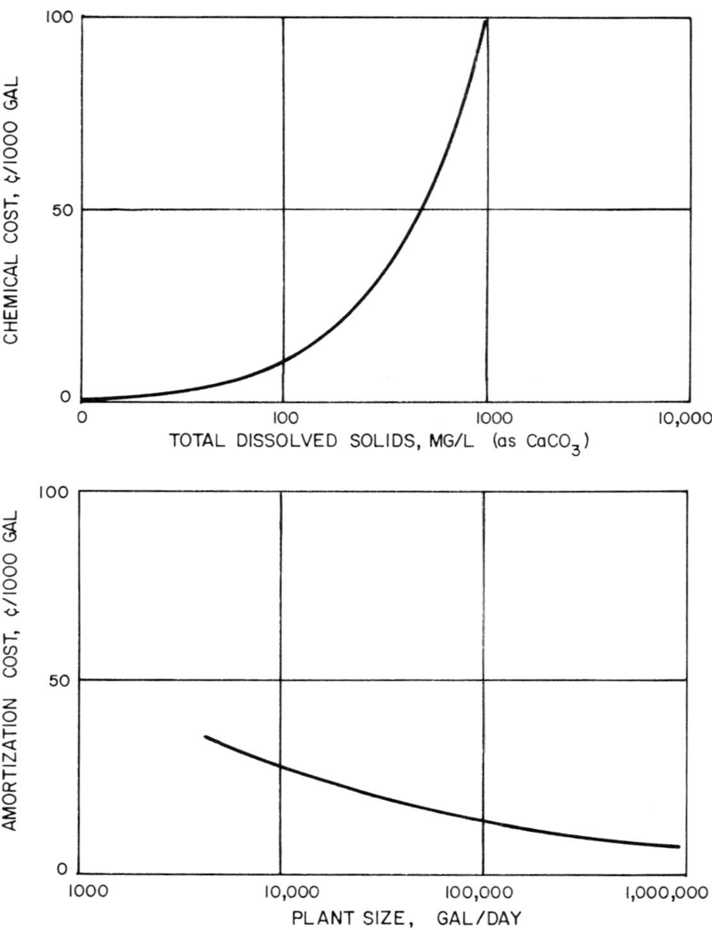

Figure 7. Chemical and amortization costs per thousand gallons for weak-base ion exhange treatment [From Ahlgren (15), courtesy of *Industrial Water Engineering*].

plant treating a 1500 mg/l chloride waste, has been estimated at \$169,000 (9). Operating costs of 26.5¢/1000 gal have been projected.

Electrodialysis

Electrodialysis has been employed for chloride treatment, particularly with saline and brackish waters in the municipal water supply industry. In a wastewater recycle operation, chlorides were reduced from 115 to 80.5 mg/l by electrodialysis (16). Other ions in the waste stream were also reduced, by the percentages shown in Table 13. Costs for this electrodialysis system (1964) are summarized in Table 14.

Table 13. Electrodialysis Removal of Ions from Wastewaters (16).

Constituent	Initial Conc. (mg/l)	Percent Reduction
Sodium	130	45
Potassium	15	40
Ammonium	20	40
Calcium	60	50
Magnesium	25	50
Chloride	115	30
Nitrate	10	60
Bicarbonate	300	60–70
Sulfate	100	40
Silica	50	0
Phosphate	25	20

Table 14. Summary of Electrodialysis Costs (16).

Cost Item	Cost (¢/1000 gal)
Amortized Capital	2.64
Operating	3.58
Maintenance	2.93
Total	9.15

Katz (17) in discussing a desalination electrodialysis plant reported chloride reduction from 122 mg/l to 58 mg/l, in addition to reduction of other ions. Estimated costs for other electrodialysis plants, achieving equivalent treatment, are summarized in Table 15. Additional operating and capital costs (6) for electrodialysis are presented in Figure 8.

Table 15. Costs of Electrodialysis Treatment of Saline Water (17).

Plant Capacity (gpd)	Capital Costs ($)	O&M Costs (¢/1000 gal)	Amortization @ 7%/yr (¢/1000 gal)	Total Water Costs (¢/1000 gal)
50,000	120,000	68.0	66.0	134.0
500,000	600,000	42.0	33.0	75.0
5,000,000	3,000,000	31.9	16.5	48.4

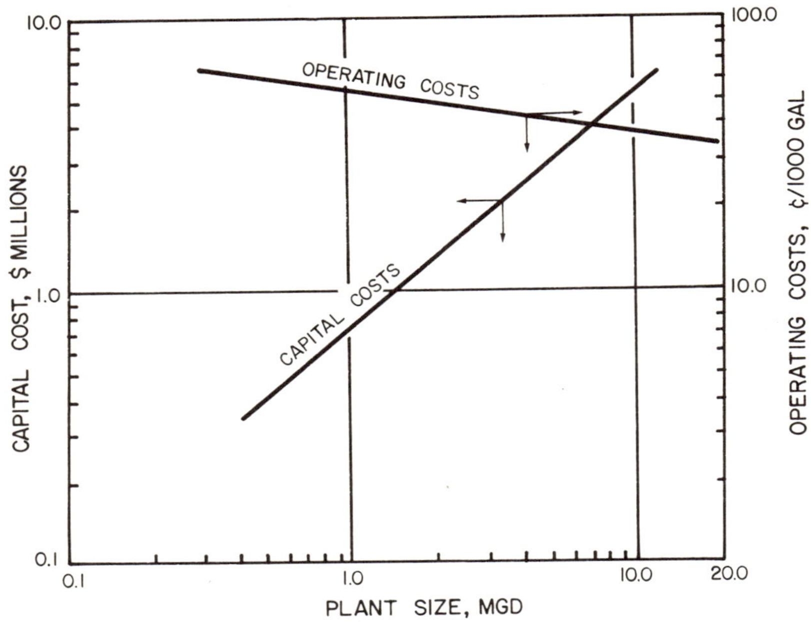

Figure 8. Capital and operating costs for electrodialysis, based on an influent TDS concentration of 3000 mg/l (6).

Table 16. Reverse Osmosis Treatment for Chloride and TDS.

Source	Volume (gpd)	Chloride (mg/l) Initial	Chloride (mg/l) Final	TDS (mg/l) Initial	TDS (mg/l) Final	Ref.
Brackish ground water	24,000	96	59	1800	155	19
Brackish ground water	150,000	351	32	2100	170	19
Brackish ground water	50,000	2020	170	4580	310	18
Sewage treatment plant effluent	10,000	260	15	756	24	20
Sewage treatment plant effluent	—	209–378	14–44	757–1170	49–92	21

Reverse Osmosis

The reverse osmosis process is effective in treating, although not particularly selective for, chloride. Typically, cellulose acetate membranes will provide 86–97% rejection of chloride, with efficiency dropping off as brine concentration increases (18). Table 16 presents treatment data from several full-scale reverse osmosis plants. Principal application of reverse osmosis has been in desalting brackish waters for domestic consumption.

Capital and operating costs for large-scale reverse osmosis systems are fairly well established. Capital costs of $470,000 for a 1.0-MGD system and $3,660,000 for a 10-MGD system have been reported in one study (21). Another study, however, estimated capital investment at $2 million and $12 million for 1.0- and 10-MGD plants respectively (22). Treatment costs are primarily dependent upon membrane life, with costs of 38–45.5¢/1000 gal for a 1-MGD plant (21). Operating costs of 45.9–55.3¢/1000 gal have been reported in a separate economic analysis of reverse osmosis (23).

SUMMARY

There are relatively few instances of treatment of chloride wastes. Intermediate levels of treatment can be achieved, however, by ion exchange, electrodialysis and reverse osmosis. For small waste volumes, costs per 1000 gallons treated are quite high. However, the economy of scale resulting from larger treatment plants results in significant reduction of unit costs. In addition to ion exchange and electrodialysis, holding basins and evaporative ponds, and deep well injection have been employed to dispose of concentrated chloride wastes. If more stringent regulations are placed on subsurface disposal, the industry's choice probably will be holding basins and evaporative ponds.

REFERENCES

1. McKee, J. E. and H. W. Wolf. *Water Quality Criteria,* California State Water Quality Control Board, Publication No. 3-A, 1963.
2. Herbert, A. J. and H. F. Berger. "A Kraft Bleach Waste Color Reduction Process Integrated with the Recovery System," *Proc. 15th Purdue Ind. Waste Conf.* **15,** 45-57 (1962).
3. *The Cost of Clean Water, Vol. III, Industrial Waste Profiles, No. I—Blast Furnaces and Steel Mills,* (Washington, D.C.: U.S. Dept. Interior, 1967).
4. Berger, M. "The Disposal of Liquid and Solid Effluents from Oil Refineries," *Proc. 21st Purdue Ind. Waste Conf.* **21,** 759-767 (1966).
5. Gloyna, E. F. and D. L. Ford. *The Characteristics and Pollutional Problems Associated with Petrochemical Wastes—Summary,* (Washington, D.C.: U.S. Dept. of the Interior, 1970).
6. *The Economics of Clean Water, Vol. III, Inorganic Chemistry Industry Profile,* (Washington, D.C.: U.S. Dept. of the Interior, 1970).

7. Jones, M. W. "Construction and Operation of a Chloride Holding Basin," *Proc. 16th Purdue Ind. Waste Conf.* **16,** 186-192 (1961).
8. Burr, D. and P. J. Byrne. "City and Industry Cooperate to Solve Brine and Waste Problems," *Publ. Works,* 46-48 (1971).
9. "Reduction of Salt Content of Food Processing Liquid Waste Effluent," U.S. EPA Report 12060 DXL 01/71, 1971.
10. Henkel, H. O. "Deep Well Disposal of Chemical Waste Water," In *Proc. 5th Ann. Sanitary and Water Resource Eng. Conf.,* Vanderbilt University, 1966.
11. *Subsurface Disposal of Industrial Wastes,* Interstate Oil Compact Commission, Oklahoma City, Oklahoma, 1968.
12. *The Cost of Clean Water, Vol. III, Industrial Waste Profiles, No. 5—Petroleum Refining,* (Washington, D.C.: U.S. Dept. of the Interior, 1967).
13. Higgins, I. R. "A Unique Process for Demineralizing Waste Water," *Ind. Eng.* **2,** 26-29 (1969).
14. Calmon, C. "Modern Ion Exchange Technology," *Ind. Water Eng.* **9,** 3:12-15 (1972).
15. Ahlgren, R. M. "Membrane vs. Resinous Ion Exchange Demineralization," *Ind. Water Eng.* **8,** 12-14 (1971).
16. Smith, J. D. and J. C. Eisenmann. "Electrodialysis in Waste Water Recycle," *Proc. 19th Purdue Ind. Waste Conf.* **19,** 738-760 (1964).
17. Katz, W. E. "Electrodialysis Saline Water Conversion for Municipal and Governmental Use," *Proc. Western Water and Power Symp.,* Los Angeles, California, 1968, pp. 113-125.
18. Cruver, J. E. "Reverse Osmosis—Where it Stands Today," *Water Sewage Works,* October, 74-78 (1973).
19. Shields, C. P. "Reverse Osmosis for Municipal Water Supply," *Water Sewage Works* **119,** 64-70, (1972).
20. Nusbaum, I., J. H. Sleigh, Jr. and S. S. Kremen. "Study and Experiments in Waste Water Reclamation by Reverse Osmosis," U.S. EPA Report 17040—05/70, 1970.
21. Cruver, J. E. and I. Nusbaum. "Application of Reverse Osmosis to Wastewater Treatment," *J. Water Poll. Cont. Fed.* **46** (2), 301-311 (1974).
22. Monti, R. P. and P. T. Silbermann. "Wastewater System Alternatives: What are They and What Cost?" *Water Wastes Eng.* **11** (6), 52-56 (1974).
23. Kremer, S. S. "The True Cost of Reverse Osmosis," *Ind. Waste* **19** (6), 24-26, (1973).

6

TREATMENT TECHNOLOGY FOR HEXAVALENT CHROMIUM

Chromium occurs in aqueous systems as both the trivalent (Cr^{+3}) and the hexavalent (Cr^{+6}) ion. Hexavalent chromium present in industrial wastes is primarily in the form of chromate ($CrO_4^{=}$) and dichromate ($Cr_2O_7^{=}$). Chromium compounds are added to cooling water to inhibit corrosion. They are employed in manufacture of inks, industrial dyes and paint pigments, in chrome tanning, aluminum anodizing and other metal cleaning, plating and electroplating operations. In the metal plating industry, automobile parts manufacturers are one of the largest producers of chromium-plated metal parts. Frequently, the major source of waste chromium is the chromic acid bath and rinse water used in such metal plating operations. Table 17 is a compilation of typical sources and concentrations of chromate wastewaters.

TREATMENT TECHNOLOGY

Reduction of hexavalent chromium from a valence state of +6 to +3, and subsequent hydroxide precipitation of the trivalent chromium ion is the most common method of hexavalent chromium dis-

Table 17. Hexavalent Chromium Wastewater Sources and Typical Concentrations.

Industrial Source	Chromium (VI) Concentration (mg/l)		Reference
	Average	Range	
Leather Tanning	40	—	1
Wood Preserving	—	0.23–1.5	2
Cooling Tower Blowdown	31.4	—	3
Cooling Tower Blowdown	—	8–10.7	4
Cooling Tower Blowdown	—	10–60	1
Bright Dip Rinse	—	1–6	5
Bright Dip Bath	—	10,000–50,000	1
Bright Dip Bath	—	20,000–75,000	6
Bright Dip Bath	—	200–600	6
Anodizing Bath	173	—	7
Anodizing Bath	—	15,000–52,000	6
Anodizing Rinse	49	—	7
Anodizing Rinse	—	30–100	6
Plating	1300	—	8
Plating	600	—	1
Plating	—	100,000–270,000	6
Plating	—	60–80	9
Electroplating	140	—	10
Electroplating	41	15–70	11
Electroplating	—	8–20.5	12

posal. The treatment of trivalent chromium is discussed in Chapter 7. To meet increasingly stringent effluent standards, some industries have turned to ion exchange to treat chromate and chromic acid wastes. Evaporative recovery of concentrated chromate and chromic acid wastes has also proved technically and economically feasible as a waste treatment alternative. The application of other processes, such as electrochemical and activated carbon adsorption techniques, is receiving increasing attention.

Reduction

The standard reduction treatment technique is to lower the waste stream pH to 3.0 or below with sulfuric acid, and convert the hexavalent chromium to trivalent chromium with a chemical reducing agent such as sulfur dioxide, sodium bisulfite, metabisulfite or hydrosulfite, or ferrous sulfate. The trivalent chromium is then removed, usually by precipitation with lime (13,14). The reduction of hexavalent to trivalent chromium is not 100% effective, and the amount of residual nonreduced hexavalent chromium depends upon the allowed time of reaction, pH of the reaction mixture, and concentration and type of reducing agent employed. The interdependence of some of these variables is shown in Figure 9. Treatment of chromic acid waste does not normally require pH adjustment, as

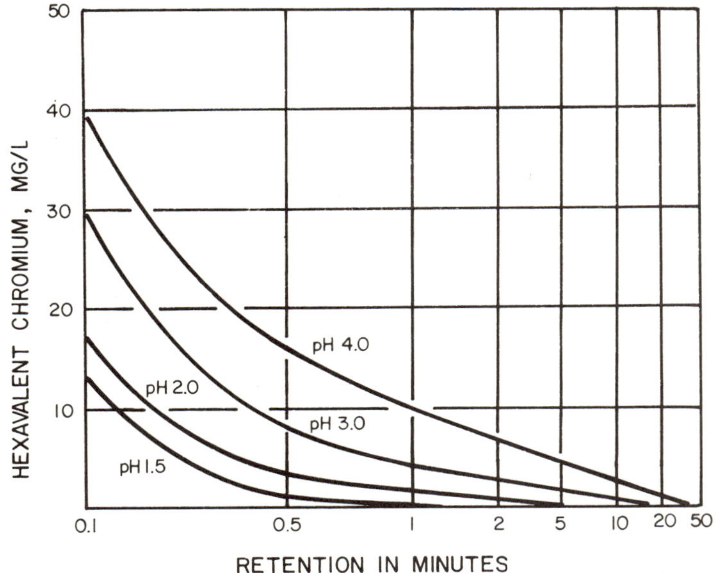

Figure 9. Effect of treatment pH and reaction time on conversion of hexavalent to trivalent chromium. [From Beevers (15), courtesy of Metal Finishing Journal.]

the pH of the waste itself is sufficiently low for the reduction reaction to proceed.

Sulfur dioxide is the most popular reducing agent used in treatment of chromium wastes, primarily because it is relatively cheap (16). Its use at a Boeing plant has been reported to treat chromium in metal finishing wastes (17). The waste was adjusted to pH 2.5 with sulfuric acid before reduction. Sulfur dioxide, used to treat chromic acid plating bath and chromic acid etch bath wastes at an average waste flow of 40 gpm, and for a 20–30 minute treatment period, yielded residual hexavalent chromium of less than 1.0 mg/l (18). Sulfur dioxide has also been used in another application to treat a waste reported to contain 1300 mg/l of hexavalent chromium. The chromate wastes were treated with sulfur dioxide at pH 2.0. Treatment detention time was approximately 90 minutes. Sulfuric acid was added to the waste, to maintain the proper pH (8). Hexavalent chromium levels of 0.01 mg/l can be achieved by reduction at pH 2.5–4.0 (19). A fully automated continuous flow chromate waste treatment system has been described that reduces hexavalent chromium levels to less than 0.05 mg/l (20). Cost of the system, including a chromic hydroxide precipitation step, was quoted as $12,000–$15,000.

The sulfur dioxide chromate reduction process has also been used to treat wood preserving wastewater (2). For an influent range of 0.23–1.5 mg/l, effluent hexavalent chromium averaged 0.1 mg/l. Cost of the treatment plant, which included chromate reduction plus precipitation of chromium, copper, fluoride and arsenic, was below $30,000. Total capital investment was approximately $50,000, including engineering, and site preparation. The daily operating cost is $10.00, plus an additional $10.00/day for laboratory analyses. At a 24 h/day wastewater flow of 20 gpm, operating costs are $0.69/1000 gal.

An unusual source of sulfur dioxide is employed at one plant, to treat hexavalent chromium in plating waste (21). Sulfur dioxide is washed out of the power house stack gas, and used as the chemical reducing agent for converting hexavalent chromium to the trivalent form. The process has been in operation since 1959. On occasions when additional reduction is necessary, sodium bisulfite is employed as a supplement. The use of sulfur dioxide for reduction imparts an oxygen demand to the waste effluent, unless the effluent is oxygenated by passing air through it (22). A required sulfur dioxide dosage of 2.0 pounds per pound of hexavalent chromium has been reported (11,15).

In the treatment of an electroplating waste containing 140 mg/l of hexavalent chromium, reduction was carried out at pH 2.5–2.8, employing sodium bisulfite as the reducing agent (10). The process was reported to reduce hexavalent chromium to concentrations of 0.7–1.0 mg/l. One attempt to use sodium bisulfite to reduce the waste from chrome plating automobile bumpers was abandoned, because of odors and corrosion hazards associated with its use (23). Sulfur dioxide is now employed in its place. By contrast, the replacement of sulfur dioxide with sodium metabisulfite as a reductant has been reported, due to industrial health problems associated with gas leakage (24). Use of sodium bisulfite to treat a chrome plating waste has been reported to yield effluent hexavalent chromium levels of 0.05–0.1 mg/l (25). Treatment costs for bisulfite reduction of $0.55 (11) and $0.80 (26) per pound of chromium treated have been given.

Two-stage chromate reduction has been used at several plants. In the first stage sulfur dioxide, sodium bisulfite or metabisulfite is employed. This is followed by a second "polishing" stage of chromate reduction with hydrazine often combined with soda-ash to simultaneously precipitate chromic hydroxide. This process has been utilized for chromium plating rinse water (12). The first stage uses sodium bisulfite at pH 2.5–4.5 and bisulfite concentration of 1000–2500 mg/l in a recirculating bath, followed by a hydrazine plus soda-ash recirculating bath at pH 7.0–8.5 and hydrazine concentration of 50–200 mg/l. This is followed by a water rinse, which overflows to a clarifier. Chromate concentration is reported reduced from 8–20.5 mg/l to less than 0.1 mg/l in the effluent (12). Treatment cost is 38.8¢/lb chromic acid treated. Similar two-stage reduction systems using hydrazine in the second stage have been reported for other plants (27,28).

The use of metabisulfite as a reducing agent has been reported for several hexavalent chromium treatment plants. One treatment plant reduced an initial chrome level of 31.4 mg/l in a cooling tower blowdown of 3600 gpm, to concentrations below 0.5 mg/l (3). Reduction was carried out at pH 2.0, obtained by addition of sulfuric acid. A metabisulfite dosage of 4.2 pounds per pound chromium was employed. This represented a 75% excess above the theoretical dosage required (13). Reductive chromate treatment of a similar blowdown waste achieved hexavalent chromium levels of 0.025–0.05 mg/l.

In another application of metabisulfite for treatment of a chromate waste from a metal plating operation, effluent hexavalent chromium levels of 0.1 mg/l or less were obtained at pH 2.0 (30). A 500-gph metabisulfite process achieved effluent hexavalent chromium levels of 0.001–0.4 mg/l, at pH 3.0 (31). The treatment plant, which also incorporated cyanide destruction, cost $16,000 (1968) for equipment plus $10,000 for the building and piping.

Sodium hydrosulfite use has been reported for chromate reduction. However, the chemical breaks down with exposure to air and must be prepared fresh daily. At one plant its use was abandoned in favor of a two-stage bisulfite-hydrazine reduction system (12).

Several reports describe the use of ferrous sulfate as a reducing agent for chromate. In three cases, treatment occurred at steel mills, where waste pickle liquor was available to use in the reduction process (5,32,33). Ferrous sulfate has been reported to have the advantage of effectiveness independent of pH (7). However, Bennet (34) claims otherwise, stating that ferrous sulfate reduction at pH

2–3 proceeds within 30 minutes to a residual hexavalent chromium level at 1 mg/l, but requires one hour at pH 4–10. It has also been pointed out that use of ferrous sulfate yields much greater quantities of sludge than are produced through use of sulfur dioxide or bisulfites, since iron precipitates will form (11,34). In addition, the use of ferrous sulfate to treat chromate wastes containing cyanide results in the formation of very stable ferrocyanide complexes, which prevent subsequent effective cyanide treatment.

For complete treatment of chromate wastes by reduction, including pH adjustment, reduction and neutralization, and removal of the trivalent chromium, treatment operating costs range from \$55–\$100/day, depending on the cost of municipal water to an industry (35). The basis for these costs is a waste flow of 100 gpm at 120 mg/l of chromium, and a 16-hour day. Cost is less than \$0.80 per pound chromium treated.

Additional cost curves are presented in Figure 10 (monthly operating costs) and Figure 11 (initial capital costs). At a flow of 70 gpm, Figure 10 indicates an operating treatment cost of ap-

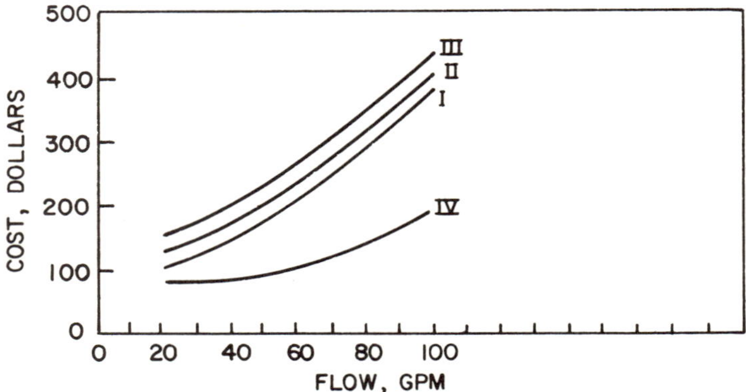

Figure 10. Monthly operating costs for two-stage chromium treatment (to trivalent), plus pH adjustment to 8.0. Depreciation plus operating costs at 50 mg/l chromium, 172 hr/month. I-no instrumentation; II-medium instrumentation; III-full instrumentation; IV-using "mist depressent" type chrome bath, no instrumentation. [From Zievers, *et al.* (36), courtesy of *Plating*.]

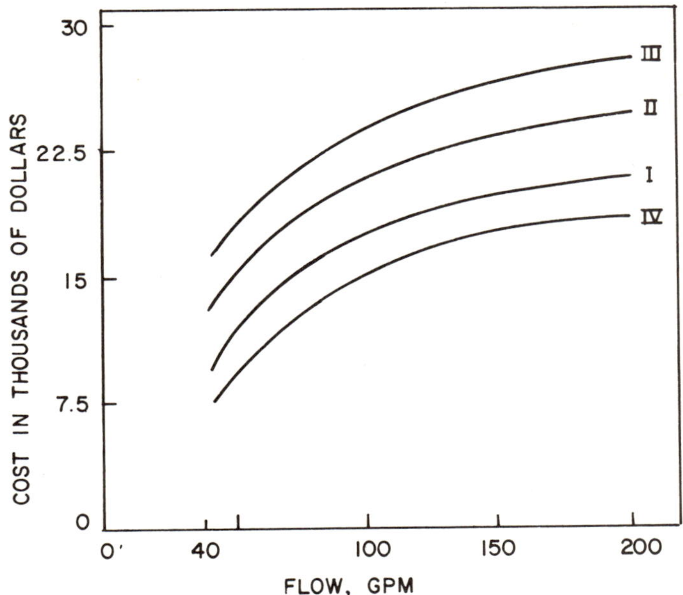

Figure 11. Initial capital costs (including installation) for chromium reduction to trivalent form. I-manual controls; II-medium instrumentation; III-full instrumentation; IV-package plant. [From Zievers, *et al.* (36), courtesy of *Plating*.]

proximately \$0.40–\$1.00/lb of chromium. These operating costs do not include the additional expense of removing the trivalent chromium. The capital costs associated with hexavalent chromium treatment are presented in Figure 11. Incremental costs associated with increased plant automation, and economy of scale resulting from treating large waste volumes are also included.

Ion Exchange

Ion exchange processes are claimed to be economical for chromium recovery and elimination of waste (37). Cation exchange can be applied to remove trivalent chromium, and anion exchange employed to remove chromate and dichromate. When the anion exchange resin is exhausted, it is regenerated (usually with sodium hydroxide), and sodium chromate is eluted from the ion exchange resin. The eluted sodium chromate can be passed through a cation exchange resin to recover purified chromic acid at concentrations as high as 6% (38). If chromic acid is not recovered, the concentrated waste yielded by the resin regeneration process must be disposed of. This may be accomplished by reduction to trivalent chromium, followed by lime precipitation. Waste treatment by ion exchange produces a reusable water, which may provide an economic edge when water costs are high.

In the ion exchange system, pH is a critical factor. At waste pH below 4, the oxidizing power of the chromic acid begins to attach the resin. At pH above 6, the ratio of chromate to dichromate in solution increases. Most anion exchange resins are less selective for dichromate, and early leakage of chromium occurs (39). Highly basic anion exchange resins preferentially remove chromate at pH 4.5–5 before phosphate or sulfate (40). Exchange capacity of 1.9 lb chromate/cu ft resin and treatment cost of 16¢/lb of chromate have been reported (40).

One significant use of ion exchange treatment is for chromate contaminated blowdown from recirculating cooling water systems (39,40). With proper pH adjustment (4.5–5) the chromate will be removed even in the presence of several hundred or thousand mg/l sulfate and chloride (39). Further, it has been reported that chromium removal from cooling tower blowdown by anion exchange pays for itself when the concentration reaches 100 mg/l chromate (40). At or above that concentration, it is a break-even proposition, and numerous plants employ ion exchange recovery systems.

Successful ion exchange treatment of a metal finishing waste has been reported, to meet a chromate effluent standard of 0.05 mg/l (41). This chromate level is equivalent to a hexavalent chromium concentration of 0.023 mg/l. The chromate and chromic acid wastes treated at this plant by ion exchange were segregated from other metal finishing wastes. Recovery of chromic acid was practiced, and resulted in a 5-year savings of $8342 above the cost that would have been required for chromium reduction and precipitation. After a year-long pilot plant test of ion exchange treatment of cooling tower blowdown containing 8–10.7 mg/l hexavalent chromium, Richardson, *et al.* (4) reported 0.0–0.90 mg/l effluent chromium levels, depending upon the period of operation between exchange resin regenerations. Upon regeneration, a chromate-rich solution was obtained and returned to the cooling system with makeup water.

A small plating shop has successfully employed ion exchange to reduce rinse water chromate levels to 0.025 mg/l (42). The concentrated eluate is treated with sodium metabisulfite to reduce chromate, followed by neutralization and chromic hydroxide precipitation.

The cost of ion exchange treatment depends upon the industrial source of the waste. Waste treatment operating costs are \$0.16–\$0.24/1000 gal of waste stream treated (22). Operating costs for ion exchange waste treatment of a waste stream of 73 gpm with a chromic acid concentration of approximately 5000 mg/l (equivalent to 2300 mg/l of hexavalent chromium), totaled \$72.24/day, with a credit of \$57.30/day resulting from chromic acid recovery and reuse of purified water (37). Net operating costs were, therefore, \$12.94/day. By contrast, reduction and precipitation would have cost \$33.19/day for that waste.

Besselievre (14) has reported ion exchange treatment of \$17/day for a chromium waste, versus an estimated reduction and precipitation cost of \$63/day. A net profit of \$10/day accrued from one ion exchange chromium recovery system, as well as a recycle of 86,500 gallons of treated wastewater to the plant (43). Yuronis has reported ion exchange costs of \$0.15/lb of chromate removed (35). He estimated capital costs for a plant treating 100 gpm and a chromium content of 50–100 mg/l as approximately \$40,000. Daily operating costs would be \$65,000, or \$0.68/1000 gal.

Electrochemical Reduction

Reduction of hexavalent to trivalent chromium has been successfully applied in an electrochemical treatment process called "cementation." The primary use of cementation in waste treatment has been to recover copper from wastewater by oxidation-reduction reactions with scrap iron (44,45). In that process, copper plates onto the scrap iron, with release of ferrous ions resulting from the oxidation half-reaction.

$$Cu^{+2} + Fe^{0} \rightarrow Cu^{0} + Fe^{+2} \quad (1)$$

However, hexavalent chromium will also react, with both elemental iron and ferrous iron, to yield trivalent chromium.

$$Cr_2O_7^{=} + 3Fe^{0} + 14H^{+} \rightarrow 2Cr^{+3} + 3Fe^{+2} + 7H_2O \quad (2)$$

$$Cr_2O_7^{=} + 6Fe^{+2} + 14H^{+} \rightarrow 2Cr^{+3} + 6Fe^{+3} + 7H_2O \quad (3)$$

The reaction requires acidic conditions and is thus pH dependent. Complete reduction of a 100 mg/l Cr^{+6} solution has been reported within 8 minutes, at pH 2.0 (45). In one full-scale application, hexavalent chromium at 2.5–52.0 mg/l was reduced to 0.01–1.40 mg/l by the cementation process. Average influent chromium was 16.3 mg/l, and average effluent 0.09 mg/l for an average chromate reduction of 99.4% (44). The process uses scrap iron, suspended in a perforated rotating drum through which the wastewater flows. The waste pH is 2–3. A similar electrochemical process, but employing an imposed electrical potential and consumable electrodes, has been reported to achieve chromate reduction to below 0.05 mg/l in cooling tower blowdown (46).

Evaporative Recovery

This process consists of evaporating metal plating rinse water, to drive off the water as vapor and thus concentrate the chromic acid for recovery and reuse. Evaporative recovery can be used on almost all process rinse water systems, with the exception of those which deteriorate with use. In evaporative recovery, not only the chromic acid but all constituents of the wastewater are retained in the concentrated product. In practice, this has been a major disadvantage of evaporative recovery since the buildup of impurities often results in defective plating such as discoloration or loss of corrosion resistance (47). Plating waste rinse water containing only a few mg/l of chromic acid can be concentrated to above 900 mg/l (48). Evaporative recovery operating costs are shown in Table 18.

Table 18. Evaporative Recovery Operating Costs (35).

Waste Flow (gpm)	Operating Costs ($/1000 gal)
0.5–2	10
2.0–5	5
5.0–10	2.5

Capital equipment costs for a continuous flow evaporative recovery system are estimated at $50,000–$60,000 (35,49).

Other Processes

Some success has been reported from pilot plant work on chromate removal by activated carbon. In one study of metal removal from municipal effluent, initial hexavalent chromium levels of 0.09–0.19 mg/l were reduced to 0.04 mg/l or less. The average effluent concentration reported was 0.017 mg/l (50). Initial hexavalent chromium levels of 5 mg/l were reduced to 0.006–0.087 mg/l following carbon adsorption (51). It appears that activated carbon may not be equally effective at higher chromate levels, however. Effluent hexavalent chromium levels of 0.05–195 mg/l have been given for carbon treatment on initial concentrations of 100–820 mg/l. Treatment efficiencies ranged from 50% to more than 99% (11).

Reverse osmosis is also reported to have application for hexavalent chromium treatment, although experience with the process is limited. Improved efficiency has been reported to result with neutralization. At pH 2.6, a high selectivity cellulose acetate membrane yielded 92.6% rejection, with 98.6% rejection at pH 7.6 (52). An additional reason for neutralization is to minimize acid hydrolysis of the reverse osmosis membrane. In pilot studies on reverse osmosis treatment of electroplating wastes, 86–98% rejection of hexavalent chromium is reported at pH 0.9–1.9, 52–99% at pH 4.4–4.7, and 91–99+% rejection at pH 5.5–6.1 (55). In addition to the pH effect, rate of flux also influences reverse osmosis performance, with generally poorer rejection at higher flux rates.

Freeze concentration of concentrated plating line waters has also been claimed economical, followed by either reuse or precipitive treatment of the concentrated slurry. In pilot plant operation, a 2500-gpd waste containing 100 mg/l of chromium yielded a product water chromium concentration of 0.225 mg/l (53).

As an alternative to chromate treatment, it has been suggested that there are instances in which proprietary trivalent chromium baths may be substituted for chromic acid bath (54). Most such formulations are organic solutions in which the trivalent chromium is at least partially stabilized by complexation with proprietary organic agents. The use of trivalent chromium formulations would eliminate the need for chromate treatment, but likely imposed severe treatment limitations on the traditional precipitation treatment used for trivalent chromium.

SUMMARY

Removal of hexavalent chromium from a waste stream has been accomplished by reduction and precipitation, ion exchange, evaporative recovery, and other techniques. Ion exchange and evaporative recovery have the advantage that the chromium may be recovered for reuse. Whether ion exchange is used to recover chemicals, or to concentrate them for further treatment, it also has the advantage of producing reusable water. Water vapor from the evaporative recovery process may be recondensed, and also reused. If waste flow rates and chromium concentrations are low (less than 50 mg/l), the total water volume reuse, and the amount of chromium involved does not appear to warrant reclamation. In such a case reduction to trivalent chromium, followed by precipitation appears most feasible. Reported treatment levels for chrome wastes are summarized in Table 19.

Table 19. Summary of Treatment Levels Reported for Hexavalent Chromium Wastes.

Treatment Process	Chromium (VI) Concentration (mg/l)		Reference
	Initial	Final	
Reduction			
Sulfur Dioxide	—	0.3–1.3	16
Sulfur Dioxide	—	1.0	18
Sulfur Dioxide	1300	"zero"	8
Sulfur Dioxide	—	0.01	19
Sulfur Dioxide	—	0.05	20
Sulfur Dioxide	0.23–1.5	0.1	2
Bisulfite	140	0.7–1.0	10
Bisulfite	—	0.05–0.1	25
Bisulfite plus Hydrazine	8–20.5	0.1	12
Metabisulfite	70	0.5	3
Metabisulfite	—	0.025–0.05	29
Metabisulfite	—	0.1	30
Metabisulfite	—	0.001–0.4	31
Ferrous Sulfate	—	1.0	34
Ion Exchange	—	0.023	41
Ion Exchange	—	0.025	42
Ion Exchange	8–10.7	0.0–0.9	4
Cementation	16.3	0.09	45
Activated Carbon*	0.09–0.19	0.017 (ave.)	50
Activated Carbon*	5.0	0.006–0.087	51
Activated Carbon*	100–820	5.05–195	11
Freeze Concentration*	100	0.225	54

*Pilot plant results.

Capital investments shown in Figure 11 appear to be substantiated by many authors. Operating and maintenance costs have been reported ranging from $0.15–$0.80/lb of chromium treated.

REFERENCES

1. Cheremisinoff, P. N. and Y. H. Habib. "Cadmium, Chromium, Lead, Mercury: A Plenary Account for Water Pollution, Part I—Occurrence Toxicity and Detection," *Water Sewage Works* **119**, 73-75, (July, 1972).
2. Teer, E. H. and L. V. Russell. "Heavy Metal Removal from Wood Preserving Wastewater," Presented at 27th Ind. Waste Conf., Purdue University, 1972.
3. Landy, J. A. "Chromate Removal at a Saudi Arabian Fertilizer Complex," *J. Water Poll. Cont. Fed.* **43**, 2292-2253 (1971).
4. Richardson, E. W., E. D. Stobbe and S. Bernstein. "Ion Exchange Traps Chromates for Reuse," *Environ. Sci. Technol.* **2**, 1006-1016 (1968).
5. Donovan, E. J., Jr. "Treatment of Wastewater for Steel Cold Finishing Mills," *Water Wastes Eng.*, (November), F22-F25 (1970).
6. Lowe, W. "The Origin and Characteristic of Toxic Wastes, with Particular Reference to the Metal Industry," *Water Poll. Cont. (London)* 270-280 (1970).
7. Germain, J., C. Vath and C. Griffin. "Solving Complex Waste Disposal Problems in the Metal Finishing Industry," Presented at Georgia Water Poll. Central Assoc. meeting, September, 1968.
8. Hansen, N. H. and W. Zabban. "Design and Operation Problems of a Continuous Automatic Plating Waste Treatment Plant at the Data Processing Division, IBM, Rochester, Minnesota," *Proc. 14th Purdue Ind. Waste Conf.*, pp. 227-249 (1959).
9. Crowle, V. "Effluent Problems as They Affect the Zinc Die-Casting and Plating—on-Plastics Industries," *Metal Finish. J.* **17** (194), 51-54 (1971).
10. Avrutskii, P. I. "Control of Chromium (VI) Concentration in Waste Waters," *Chem. Abstr.* **70**, 206-207 (1969).
11. "An Investigation of Techniques for Removal of Chromium for Electroplating Wastes," U.S. EPA Publication 12010 EIE 13/71 (1971).
12. Martin, J. J., Jr. "Chemical Treatment of Plating Waste for Removal of Heavy Metals," U.S. Environmental Protection Agency, EPA-R2-73-044 (May, 1973).
13. MacDougall, H. "Waste Disposal at a Steel Plant: Treatment of Sheet and Tin Mill Wastes," *ASCE Proc.* Separate No. 493 (1954).
14. Besselievre, E. B. *The Treatment of Industrial Wastes* (New York: McGraw-Hill Book Co., 1969).
15. Beevers, M. "Chlorine and Sulfur Dioxide in the Treatment of Cyanide and Chromium Wastes," *Metal Finish. J.* **18** (211), 232-235 (1972).

16. Stone, E. H. F. "Treament of Non-Ferrous Metal Process Wastes," *Metal Finish. J.* **18** (212), 280-290, (1972).
17. Halse, B. T., R. P. Selim and G. E. Summers. "Control of Metal Finishing Wastes Using ORP," *J. Water Poll. Cont. Fed.* **32,** 975-981 (1960).
18. Shink, C. A. "Plating Wastes: A Simplified Approach to Treatment," *Plating* **55,** 1302-1305 (1968).
19. Curry, N. A. "Philosophy and Methodology of Metallic Waste Treatment," Presented at 27th Ind. Waste Conf., Purdue University, 1972.
20. Lavin, D. and C. Lavin. "A Practical Approach to Water Pollution Control for the Plating Company," Presented at 27th Ind. Waste Conf., Purdue University, 1972.
21. Fisco, R. "Plating and Industrial Waste Treatment at the Expanded Plant of Fisher Body, Elyria, Ohio," *Proc. 25th Purdue Ind. Waste Conf.* 1970.
22. Lacy, W. J. and A. Cywin. "The Federal Water Pollution Control Administration's Research and Development Program: Industrial Pollution Control," *Plating* **55,** 1299-1301 (1968).
23. Avrutskii, P. I. "Liquid SO_2 in New Reduction Role," *Ind. Eng.* **53,** 29A-30A (1961).
24. Kelsey, G. D. and W. E. Seeds. "Fully Automated Effluent Treatment Plant," *Metal Finish. J.* **18** (211), 246-248 (1972).
25. "Effluent Treatment at B.S.R.," *Metal Finish. J.* **17**, 201 (1971).
26. Coulter, K. R. "Pollution Control and the Plating Industry," *Plating* **59,** 1197-1202 (1970).
27. Crowle, V. A. "Effluent Treatment and Materials Recovery from the Metal Finishing Industry Using the Integrated Method of Treatment," *Water Poll. Cont.* 636-645 (1972).
28. "Zinc Plating Effluent—Treatment and Recovery," *Effluent Water Treat. J.* **12** (10), 543 (1973).
29. Cheremisinoff, P. N. and Y. H. Habib. "Cadmium, Chromium, Lead and Mercury: A Plenary Account for Water Pollution," *Water Sewage Works* **119,** 45-51 (1972).
30. Werner, H. W. "Treatment of Metal Finishing Wastes by Robertshaw Controls Company," presented at 27th Ind. Waste Conf., Purdue University, 1972.
31. O'Neill, F. "Facing up to Pollution," *Plating* **57,** 1211-1213 (1970).
32. Heynike, J. J. C. and F. V. K. von Reiche. "Water and Pollution Control in the Iron and Steel Industry, with Special Reference to the South African Iron and Steel Industrial Corporation," *Water Poll. Cont. (London)* 569-573 (1969).
33. Stoner, L. B. "Waste Treatment Facilities for Jones and Laughlin Steel Corporation, Hennepin Works," *Proc. 26th Ind. Waste Conf.* **26,** 761-765 (1971).

34. Bennett, J. R. "The Treatment of Effluents from Metal Cleaning and Finishing Processes," *Metal Finishing J.* **18** (212), 272-276 (1972).
35. Yuronis, D. "Metal Finishing Waste Treatment: Comparative Economics," *Plating* **55,** 1071-1074 (1968).
36. Zievers, J. F., R. W. Crain and F. G. Barclay. "Waste Treatment in Metal Finishing: U.S. and European Practices," *Plating* **55,** 1171-1179 (1968).
37. Ross, R. D. (Ed.), *Industrial Waste Disposal,* (New York: Von Nostrand Reinhold Co., 1968).
38. Dvorin, R. "Water and Waste Treatment—A Review of Methods for the Metal Finishing Industry," *Metal Finishing Guidebook—Directory* (1960).
39. Anderson, R. E. "Some Examples of the Concentration of Trace Heavy Metals with Ion Exchange Resins," *Proc. Traces of Heavy Metals in Water-Removal Processes and Monitoring,* U.S. Environmental Protection Agency, Report EPA—902/9-74-001 (1974).
40. Calmon, C. "Trace Heavy Metal Removal by Ion Exchange," *Proc. Traces of Heavy Metals in Water-Removal Processes and Monitoring,* U.S. Environmental Protection Agency, Report EPA—902/9-74-001 (1974).
41. Rothstein, S. "Five Years of Ion Exchange," *Plating* **45,** 835-841 (1958).
42. "Compact Ion-Exchange System Works for 'Small' Plater," *Ind. Waste* **19** (1), 22-23 (1973).
43. Keating, R. J. and J. H. Duff. "Plating Waste Solutions: Recovery or Disposal," *Proc. 10th Purdue Ind. Waste Conf.* (1955).
44. Joster, T. L. and T. H. Taylor. "Industrial Waste Treatment at Scovill Manufacturing Company," Presented at 28th Ind. Waste Conf., Purdue University, 1973.
45. Chase, O. P. "Metallic Recovery from Waste Waters Utilizing Cementation," U.S. Environmental Protection Agency, Report EPA—670/2-74-008 (1974).
46. Duffey, J. G. Andco Environmental Processes, Inc., Personal Communication, July 24, 1974.
47. Pinner, R. "The Future of Effluent Treatment in Metal Finishing: Part II," *Metal Finish. J.* **19** (218), 82-84 (1973).
48. Culotta, J. M. and W. F. Swanton. "Case Histories of Plating Waste Recovery Systems," Presented at 56th Annual Conf. Amer. Electroplaters Soc., Detroit, 1969.
49. Culotta, J. M. and W. F. Swanton. "Recovery of Plating Wastes: Selection of Lowest Cost Evaporator," *Plating* **57,** 1221-1223 (1970).
50. Argo, D. G. and G. L. Culp. "Heavy Metals Removal in Wastewater Treatment Processes: Part 2—Pilot Plant Operation," *Water Sewage Works* **119** (9), 128-132 (1972).

51. Maruyama, T., S. A. Hannah and J. M. Cohen. "Removal of Heavy Metals by Physical and Chemical Treatment Processes," Presented at 45th Annual Conf., Water Poll. Cont. Fed., 1972.
52. Cruver, J. E. "Reverse Osmosis—Where It Stands Today," *Water Sewage Works,* 74-77 (October, 1973).
53. Donnelly, R. G., R. L. Goldsmith, K. J. McNulty and M. Tan. "Reverse Osmosis Treatment of Electroplating Wastes," *Plating* **61** (5), 432-442 (1974).
54. Cambell, R. J. and D. K. Emmerman. "Freezing and Recycling of Plating Rinsewater," *Ind. Water Eng.* **9** (4), 38-39 (1972).
55. Silman, H. "The Substitution of Toxic Materials," *Plating* **61** (4), 332-336 (1974).

7

TREATMENT TECHNOLOGY FOR TRIVALENT CHROMIUM

Chromium in industrial wastes occurs predominantly in the hexavalent form, as chromate ($CrO_4^=$) and dichromate ($Cr_2O_7^=$). Industrial sources and treatment of these ions is discussed in Chapter 6. Hexavalent chromium treatment normally involves reduction to the trivalent (Cr^{+3}) form, prior to removing the chromium from the industrial waste. Thus trivalent chromium in industrial waste may result from one step of the waste treatment process itself, that of chemical reduction of hexavalent chromium. Industries that employ trivalent chromium directly in manufacturing processes include glass, ceramics, photography, inorganic pigments, textile dyeing, and animal-glue manufacture. Proprietary trivalent chromium plating solutions have been developed as substitutes for chromic acid plating baths, although their use is limited (1).

In addition to the predominant hexavalent form, some few mg/l of trivalent chromium might be encountered in chromic acid plating wastes, even prior to chemical reduction of Cr^{+6}. For example, one plant waste contained hexavalent chromium concentrations ranging from 0.0–18.0 mg/l and total chromium concentrations of 2.9–31.8

mg/l (2). By contrast, a trivalent chromium concentration in this one waste stream of 2.9–13.9 mg/l is indicated. For the same plant, an acid bath waste stream contained hexavalent chromium at 122–270 mg/l and trivalent chromium levels of 37–282 mg/l. Evaluation of plating rinse data from other plants indicates trivalent chromium levels of 2.5–15 mg/l prior to treatment (3,4).

TREATMENT TECHNOLOGY

Trivalent chromium may be removed by precipitation as the hydroxide upon addition of lime or caustic, or may be concentrated by ion exchange and recovered. In one instance, precipitation as chromic sulfide has been reported (5). The high cost of chromium as an industrial material indicates that ion exchange and recovery should be seriously considered as a waste treatment process. However, at this time, precipitation of chromic hydroxide and disposal of the resulting sludge remains the most common practice.

Precipitation

Trivalent chromium can be removed as insoluble chromic hydroxide, $Cr(OH)_3$, by precipitation with caustic soda or lime (6). The chromic hydroxide sludge formed by the process frequently is disposed by landfill. The precipitation process is most effective at pH 8.5–9.5, due to the low solubility of chromic hydroxide in that range (Figure 12), although best removal has been reported in one case at pH 12.2 for trivalent chromium in an animal glue manufacturing waste (Table 20). Because a typical metal-bearing waste will include other metal ions as well as chromium to be removed by precipitation, and because the optimum pH of precipitation varies for different metal hydroxides, an average waste stream pH of about 8 often seems to achieve the best overall results on a waste containing several metal ions (9). Thus, much of chromium treatment reported in the literature occurred at pH 7–8.

Soda ash has been used to precipitate trivalent chromium (6,10), although the most commonly employed chemical for precipitation is lime. Reduction of the chromium content of one electroplating waste from 140 to 1.0 mg/l was achieved by neutralization and precipitation with lime, at pH 7–8 (11). In waste treatment at an IBM plant, hexavalent chromium (at 1300 mg/l) in one waste stream was converted to trivalent chromium and precipitated with lime to achieve an effluent containing no hexavalent and 0.06 mg/l trivalent chromium (12). A coagulant aid was employed to improve the

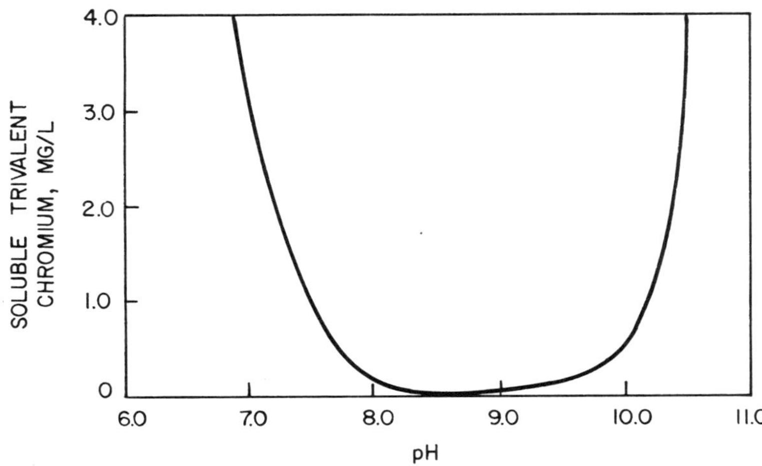

Figure 12. Effect of pH on solubility of trivalent chromium. [From Parsons (7), courtesy of National Lime Association.]

Table 20. Trivalent Chromium Precipitation Treatment (8).

pH	Chromium (mg/l)		Percent Removal
	Initial	Final	
4.5	650	500	23
8.8	650	18	97
12.2	650	0.3	99.9+

precipitation-sedimentation removal of chromic hydroxide. This coagulant aid undoubtedly accounted for the high degree of treatment achieved, since the influent to the sedimentation basin contained 1000–2000 mg/l suspended solids, while the effluent routinely contained less than 10 mg/l. Thus 99–99.5% suspended solids removal was achieved by employing the coagulant aid. Equally good effluent hydrous metallic suspended solids levels were reported in another case where a coagulant aid was also employed to improve chromic hydroxide precipitate removal (13).

Filtration of effluent from the postprecipitation sedimentation basin has been employed to remove the small, nonsettleable chromic hydroxide particles not readily removable by simple sedimentation. In one instance the use of filtration significantly improved the effluent quality of a chromium-bearing waste from a British metal alloy plant (14). Hexavalent chromium as sodium dichromate was employed to produce a bright finish on copper alloy products. The plating was carried out in a mixture of the sodium dichromate and sulfuric acid. Waste chromate from the process, at 7400 mg/l, was reduced with sulfur dioxide to trivalent chromium, and precipitated at pH 8.5 with lime. Hydroxides of copper, zinc and nickel also precipitated in the process. The metal hydroxides were removed by sedimentation. During a 2-week surveillance period, chromium content in the settling basin effluent was found to range from 1.3–4.6 mg/l. Passage of this effluent through a sand filter to remove the finely divided chromic hydroxide particles carried over reduced the effluent chromium concentrations to 0.3–1.3 mg/l. This residual should principally be in the soluble Cr^{+6} form, rather than as trivalent chromium.

One process described in the literature involves chromate reduction by sodium metabisulfite to below 0.1 mg/l, followed by waste neutralization with lime to pH 9.0, precipitation of chromic plus other metallic hydroxides, and settling (15). Trivalent chromium levels below 0.2 mg/l can be achieved by such treatment (16). Lime precipitation of a dilute wood preserving wastewater, containing trivalent chromium up to 2.2 mg/l, yielded an average effluent of 0.02 mg/l (17). Solids separation was by settling.

The use of reduction and lime precipitation to treat chromium-bearing waste from a General Electric appliance plant has been reported (2). Sedimentation of the precipitate was improved by addition of an anionic polyelectrolyte. Plant effluent was reported to contain no hexavalent chromium and an average 0.75 mg/l trivalent chromium.

Large quantities of sludge are produced in the precipitation of metal hydroxides from industrial wastes. Typically these sludges settle rapidly but do not compact very well, and contain only 2–3% solids. Various methods of sludge concentration and disposal are presented in Table 21, as reported for typical industrial processes.

Table 21. Metal Hydroxide Sludge Concentration and Disposal Technique (18).

Method	Comments
Lagooning	Requires large areas of land. Might contaminate ground waters unless properly lined.
Vacuum Filtration	Equipment and operating costs high. Discharge cake dry enough for haulage.
Pressure Filtration	Equipment and operating costs high. Requires proper conditioning of sludge. Dry cake.
Centrifugation	Method of solids removal for large volumes and large particles.

Table 22. Treatment of Metal Hydroxide Sludges.

Initial Sludge Solids (mg/l)	Sludge Dewatering Process	Final Sludge Solids (mg/l)	Ultimate Sludge Disposal	Reference
—	Vacuum Filtration	—	Landfill	2,6,15
30,000–60,000	Vacuum Filtration	100,000–120,000	City Dump	12
100,000	Vacuum Filtration	290,000	?	14
—	Vacuum Filtration	200,000–250,000	?	19
30,000–50,000	None	30,000–50,000	Landfill	20

The filtrate from sludge dewatering is normally returned to the original sludge sedimentation basin for further treatment.

In one case where vacuum filtration was not necessary, sludge from the sedimentation basin was held for short periods of time in sludge storage tanks (12). The sludge stored in the holding tank thickened (or compacted) to a suspended solids concentration of 10–12%, equivalent to that obtainable by vacuum filtration. The thickening was thought due to the application of a coagulant aid in the sedimentation basin, although another possible explanation is that the chromic hydroxide sludge aged and underwent transformation to the more dense chromic oxide form.

Treatment data on chromic hydroxide precipitation of aircraft engine manufacturing wastes has been published, which indicates that concentrated wastes yield most dense sludge, and dilute wastes yield a more voluminous sludge (21). Results from chromic hydroxide treatment suggest that each volume of dilution water added to an equal initial volume of concentrated trivalent chromium waste increases the final settled sludge volume by 49%. This would indicate a major benefit in final sludge handling for treating concentrated rather than diluted weak wastes.

The treatment costs associated with trivalent chromium removal by precipitation are due to the lime precipitation process and the expense of sludge collection and disposal. Yuronis (22) estimates a lime cost of approximately $0.15/lb of trivalent chromium precipitated. For example, a waste containing 1000 mg/l Cr^{+3} would have a lime cost of $1.25/1000 gal. Cost of complete treatment from the hexavalent chromium form would include reduction cost, precipitation cost, and sludge handling, and disposal costs. Reduction costs are approximately $1.00/1000 gal.

Capital costs of $12,000–$15,000 have been quoted for typical off-the-shelf automated and continuous-flow chromium treatment plants, which incorporate chromate reduction, pH adjustment and chromic hydroxide precipitation (15). Operating costs ranging from

\$0.45–\$1.90/1000 gal of chromium wastewater treated, depending upon the chemicals used, have been reported (18). This includes reducing as well as precipitating chemicals.

There is a growing emphasis on filtration of metal hydroxide sludges, due to more stringent enforcement of effluent suspended solids standards. Figures 13 and 14 present chromic hydroxide sludge filtration operating equipment costs. Figure 13 presents operating expenses as a function of sludge flow rate in gallons per minute versus monthly operating cost in thousands of dollars. Three operating conditions are presented in Figure 13, based upon the use of bodyfeed and/or flocculating agents to improve filtration. Bodyfeed cost was based upon use of diatomaceous earth as a filter aid. The filterability of chromium sludge precipitated by caustic is poorer than that of sludge precipitated with lime, and the former sludge frequently requires a bodyfeed filter-aid to improve dewatering characteristics, while the latter sludge does not. Figure 14 shows equipment costs in thousands of dollars, versus sludge flow in gallons per minute. The economy of scale associated with high waste flows is evident.

Ion Exchange

Relatively little information is available in the technical literature on ion exchange treatment of trivalent chromium. This is partly because most industrial waste chromium exists in the hexavalent form, and if recovery is preferred, direct ion exchange of the hexavalent chromate or dichromate is employed. Ion exchange is dependent upon the electron charge on an ion. Trivalent chromium has a +3 charge, and hexavalent chromium in its usual form of chromate or dichromate has a −2 charge. Thus, different ion exchange resins must be employed to recover each form of chromium. An anion type of resin is required to capture negatively charged forms such as the hexavalent chromate and dichromate ions. A cationic type of resin is employed for positively charged (*e.g.*, Cr^{+3}) forms.

Two reasons have been suggested to recover trivalent chromium by ion exchange (23). The first is the passage of concentrated chromic acid baths through a cation resin to remove metallic contaminants such as iron, aluminum and trivalent chromium from the chromic acid. This process would purify the chromic acid solution, which could then be reused. One difficulty with this application is that chromic acid, at the concentrations typically used in plating

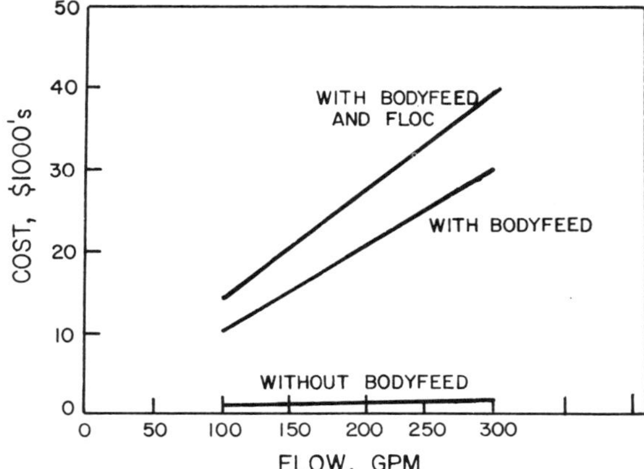

Figure 13. Monthly operating plus depreciation costs (12-year straight line) for wastewater filtration, 720 hr/month. [From Zievers, *et al.* (9), courtesy of *Plating.*]

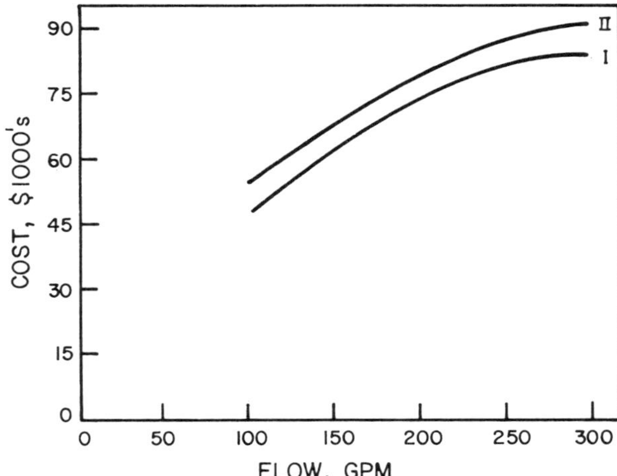

Figure 14. Initial capital costs for filtration equipment, including installation. I-manual operation; II-semi automatic operation. [From Zievers, *et al.* (9), courtesy of *Plating.*]

baths (250–400 g/l), will attack and destroy exchange resin. Experience has indicated that the maximum acceptable concentration of chromic acid is approximately 250 g/l (24). Therefore, chromic acid solutions above this concentration must be diluted before ion exchange treatment. After ion exchange, reconcentration of the purified chromic acid by evaporation in glass-lined vessels is common (24).

The second reason is for the complete purification of a plating rinse water, in which a rather dilute waste would be passed through both an anionic and a cationic resin to remove chromium species of both positive and negative charge (23). The treated rinse water could then be recycled. No capital-cost figures are available for ion exchange treatment of trivalent chromium. Lacy and Cywin (25) have reported operating costs for ion exchange treatment of \$0.16–0.25/1000 gal. Ross (23) has reported daily operating costs for ion exchange treatment of trivalent chromium as approximately \$28/day for a waste flow of 73 gal/min. Assuming a 16-hour day, operating costs are estimated at \$0.40/1000 gal.

SUMMARY

Most trivalent chromium in industrial waste results from reduction of hexavalent chromium as a preliminary step in waste treatment. This trivalent chromium is most often treated with lime to precipitate chromic hydroxide. Table 23 summarizes results of precipitation treatment. The chromic hydroxide sludge is dewatered by vacuum filtration, or disposed of directly to landfill or otherwise. Treated effluent trivalent chromium concentrations of 0.02 mg/l have been reported (17). Under special circumstances, ion exchange removal of trivalent chromium has been employed. Ion exchange is characterized by effecting complete Cr^{+3} removal, until the exchange capacity of the resin is exhausted. This exchange capacity may be recovered by regenerating the resin.

Table 23. Summary of Trivalent Chromium Treatment.

Method	pH	Chromium (mg/l)		Reference
		Initial	Final	
Precipitation	—	—	0.75	2
Precipitation	8.8	650	18	2
Precipitation	12.2	650	0.3	8
Precipitation	7–8	140	1.0	11
Precipitation	—	1300	0.06	12
Precipitation	8.5	7400	1.3–4.6	14
with sand filtration		7400	0.3–1.3	14
Precipitation		2.2	0.02	17

REFERENCES

1. Silman, H. "The Substitution of Toxic Materials," *Plating* **61** (4), 332-336 (1974).
2. Anderson, J. S. and E. H. Iobst, Jr. "Case History of Wastewater Treatment in a General Electric Appliance Plant," *J. Water Poll. Cont. Fed.* **10,** 1786-1795 (1968).
3. Germain, J. E., C. A. Vath and C. F. Griffin. "Solving Complex Waste Disposal Problems in the Metal Finishing Industry," Presented at Georgia Water Poll. Cont. Assoc. meeting, September, 1968.
4. Martin, J. J., Jr. "Chemical Treatment of Plating Waste for Removal of Heavy Metals," U.S. Environmental Protection Agency, Report EPA—R2-72-044 (1973).
5. Eldridge, J. D. and T. L. Judell. "Recovery of Materials from Process Effluents," In *Solid Waste Treatment and Disposal,* N.Y. Kirov, Editor, (Ann Arbor, Michigan: Ann Arbor Science Publishers Inc., 1972).
6. Besselievre, E. B. *The Treatment of Industrial Wastes,* (New York: McGraw-Hill Book Co., 1969).
7. Parsons, W. A. *Chemical Treatment of Sewage and Industrial Wastes,* (Washington, D.C.: National Lime Association, 1965).
8. Wang, L. K., R. P. Leonard, J. G. Michalouic and D. W. Goupil. "Glue Treatment—Pick a Way," *Water Waste Eng.* **10** (9), 20-24 (1973).
9. Zievers, J. F., R. W. Crain and F. G. Barclay. "Waste Treatment in Metal Finishings, U.S. and European Practices," *Plating* **55,** 1171-1179 (1968).
10. "Liquid SO_2 in New Reduction Role," *Ind. Eng. Chem.* **53,** 29A-30A (1961).
11. Aurutskii, P. I. "Control of Chromium (VI) Concentration in Waste Waters," *Chem. Abstr* **70,** 206-207 (1969).
12. Hansen, N. H. and W. Zabben. "Design and Operation Problems of a Continuous Automatic Plating Waste Treatment Plant at the Data Processing Division, IBM, Rochester, Minnesota," *Proc. 14th Purdue Ind. Waste Conf.,* 227-249 (1959).
13. O'Connor, S. F., B. W. Montjor, Jr. and N. S. Chamberlin. "Western Electric Builds Modern Plant for Treating Metal Finishing Wastes," *Water Wastes Eng./Ind.,* D16-D19 (1962).
14. Stone, E. H. F. "Treatment of Non-Ferrous Metal Process Waste at Kynoch Works, Birmingham, England," *Proc. 22nd Ind. Waste Conf.* 848-865 (1967).
15. Werner, H. W. "Treatment of Metal Finishing Wastes by Robertshaw Control Company," Presented at 27th Ind. Waste Conf., Purdue University, 1972.
16. Curry, N. A. "Philosophy and Methodology of Metallic Waste Treatment," Presented at 27th Ind. Waste Conf., Purdue University, 1972.

17. Teer, E. H. and L. V. Russell. "Heavy Metals Removal from Wood Preserving Wastewater," Presented at 27th Ind. Waste Conf., Purdue University, 1972.
18. Anderson, J. S., J. M. Phillips and C. B. Schriver. "Water Contamination by Certain Heavy Metals," Presented at 44th Annual Water Poll. Cont. Fed. Conf., 1971.
19. Kelsey, G. D. and W. E. Seeds. "Fully Automated Effluent Treatment Plant," *Metal Finish. J.* **18** (211), 246-248 (1972).
20. Fisco, R. "Plating and Industrial Waste Treatment at the Expanded Plant of Fisher Body, Elyria, Ohio," *Proc. 25th Purdue Ind. Waste Conf.* (1970).
21. Chalmers, R. K. "Treatment of Wastes from Metal Finishing and Engineering Industries," In *Program in Water Technology, Volume I, Applications of New Concepts of Physical-Chemical Wastewater Treatment,* W. W. Eckenfelder and L. K. Cecil, Editors, (New York: Pergamon Press, Inc., 1974).
22. Yuronis, D. "Metal Fnishing Waste Treatment: Comparative Economics," *Plating* **55**, 1071-1074 (1968).
23. Ross, R. D. (Ed.), *Industrial Waste Disposal,* (New York: van Nostrand Reinhold Book Co., 1968).
24. Thompson, J. and V. J. Miller. "Role of Ion Exchange in Treatment of Metal Finishing Wastes," *Plating* **58**, 809-812 (1971).
25. Lacy, W. J. and A. Cywin. "The Federal Water Pollution Control Administration's Research and Development Program: Industrial Pollution Control," *Plating* **55**, 1299-1301 (1968).

8

TREATMENT TECHNOLOGY FOR COPPER

INDUSTRIAL SOURCES AND LEVELS

A principal source of copper in industrial waste streams is metal cleaning and plating baths, and rinses, as brass and copper metal-working requires periodic oxide removal by immersing the metal in strong acid baths. Solution adhering to the cleaned metal surface, referred to as "dragout," is rinsed from the metal and contaminates the waste rinse water. Similarly, metal parts undergoing copper or brass plating drag out some of the concentrated plating solution. Copper concentration in the plating bath depends upon the bath type, and may range from 3,000–50,000 mg/l.

For a given bath, the rinse water concentration will be a function of many factors including drainage time over the bath, shape of the part and total surface area, method of rinsing and the rate of rinse water flow. Rinse water copper concentration may range from 0.02–1.0% of the process bath concentration (1,2).

Jewelry manufacturers employ copper plating either directly or as a base metal for silver and other precious metals. Copper is also employed in the alkaline Bemberg rayon process, as cuproammonium salts. Copper-bearing mining wastes and acid mine drainage contribute significant quantities of dissolved copper to waste streams. Additional potential sources of copper-bearing waste include pulp, paper, and paperboard mills, wood preserving, fertilizer manufacturing, petroleum refining, paints and pigments, basic steel works and foundaries, nonferrous metalworks and foundries, and motor vehicle and aircraft plating and finishing (3). Process wastewater concentrations are summarized in Table 24.

TREATMENT TECHNOLOGY

As with most heavy-metal wastes, treatment processes employed for reduction of soluble copper in wastewaters may involve precipitation and disposal of resulting solids, or recovery processes such as ion exchange, evaporation and electrolysis. The value of copper metal frequently makes recovery processes more attractive than for iron or zinc. In precipitation treatment, a copper-rich sludge may have significant value for copper recovery by an outside processor.

The importance of good housekeeping and process modification in reducing overall treatment requirements, particularly in metal processing and plating operations, has been shown in many instances (22-26). Principally, modification of rinsing techniques by installing cascading or counter-current rinse systems increases the contaminant concentration, while reducing the total wastewater volume. Recovery processes may become economically and technologically feasible through such modifications. Ion exchange is an appropriate treatment method for wastewaters containing copper at less than 200 mg/l; precipitation is applicable for copper levels of 1.0–1000 mg/l, and electrolytic recovery is advantageous for copper treatment at wastewater concentrations above 10,000 mg/l (3).

Precipitation

The standard treatment method for copper and most other heavy metals is precipitation as the relatively insoluble hydroxide at alkaline pH, or occasionally as the sulfide. Generally, hydroxide precipitation is accomplished by lime addition to an acidic wastewater. Lime is preferred to the more expensive sodium or potassium hydroxide, as it costs significantly less. In the presence of high sulfate, considerable calcium sulfate precipitate may form, and recovery value of the sludge copper may dictate use of a more expensive base to obtain a purer sludge.

Cupric oxide, which forms from the hydroxide, has minimal solu-

Table 24. Concentrations of Copper in Industrial Process Wastewaters.

Process	Copper Concentration (mg/l)	Reference
Plating Rinse	20–120	4
Plating Rinse	0–7.9	5
Plating Rinse	20 (ave.)	6
Plating Rinse	5.2–41	7
Plating	6.4–88	8
Plating	2.0–36.0	9
Plating	20–30	10
Plating	10–15	10
Plating	3–8	10
Plating	11.4	11
Appliance Manufacturing		11
Spent Acids	0.6–11.0	
Alkaline Wastes	0–1.0	
Automobile Heater Production	24–33 (28 ave.)	12
Silver Plating		13
Silver Bearing	3–900 (12 ave.)	
Acid Wastes	30–590 (135 ave.)	
Alkaline Wastes	3.2–19 (6.1 ave.)	
Brass Plating		13
Pickling Bath Wastes	4.0–23	
Bright Dip Wastes	7.0–44	
Plating Wastes	2.8–7.8 (4.5 ave.)	13
Pickling Wastes	0.4–2.2 (1.0 ave.)	13
Brass Dip	2–6	5
Brass Mill Rinse	4.4–8.5	14
Brass Mill Rinse		15
Tube Mill	74	
Rod and Wire Mill	888	
Brass Mill Bichromate Pickle		15
Tube Mill	13.1	
Rod and Wire Mill	27.4	
Rolling Mill	12.2	
Copper Rinse	13–74	
Brass Mill Rinse	4.5	
Brass and Copper Wire Mill	75–124	16
Brass and Copper Pickle	60–9	16
Brass and Copper Bright Dip	20–35	16
Copper Mill Rinse	19–74	14
Copper Tube Mill	70 (ave.)	17
Copper Wire Mill	800 (ave.)	17
Copper Ore Extraction	0.28–0.33	18
Gold Ore Extraction	20	19
Acid Mine Drainage	3.2	20
Acid Mine Drainage	3.9	20
Acid Mine Drainage	0.12	20
Acid Mine Drainage	51.6–128.0	21

bility between pH 9.0 and 10.3 (27). A minimal solubility of 0.01 mg/l has been reported in that pH range, from laboratory studies (28). This value corresponds closely to the theoretical minimum effluent level which could be achieved by precipitation. Theoretical levels are seldom attained in actual practice, due to poor separation of colloidal precipitates, slow reaction rates, pH fluctuations and the influence of other ions in solution. Reported treatment levels achieved by precipitation in full-scale industrial treatment operations are presented in Table 25. These data demonstrate the impact of effective solids separation.

Maruyama, *et al.* (39) studied copper removal from municipal wastewater enriched to 5.0 mg/l. Various combinations of lime and ferrous sulfate addition were examined for treatment effectiveness, in a 4-gpm pilot plant. After chemical addition, flocculation and settling, the effluent was sand filtered and passed through activated carbon columns. Treatment levels achieved are summarized in Table 26. It is noteworthy that best copper removal occurred at

Table 25. Summary of Effluent Copper Concentrations after Precipitation Treatment.

Source and Treatment	Initial Copper conc. (mg/l)	Final Copper conc. (mg/l)	Reference
Metal Processing (Lime)	204–385	0.5	29
Nonferrous Metal Processing (Lime)	—	0.2–2.3 (prior to sand filtration)	30,31
Metal Processing (Lime)	—	1.4–7.8 (prior to sand filtration) 0.0–0.5 (after sand filtration)	32
Electroplating (caustic, Soda Ash + Hydrazine)	6.0–15.5	0.09–0.25 (sol.) 0.30–0.45 (tot.)	33
Machine Plating (Lime + coagulant)	—	2.2	34
Metal Finishing (Lime)	—	0–1.2 (ave. 0.19)	35
Brass Mill (Lime)	10–20	1–2	17
Plating	—	0.02–0.2	36
Plating (CN oxidation, Cr reduction, neutralization)	11.4	2.0	37
Wood Preserving (Lime)	0.25–1.1 (range)	0.1–0.35	38
Brass Mill (Hydrazine + NaOH)	75–124	0.25–0.85	16
Silver Plating (CN oxidation, Lime, $FeCl_3$	30 (ave.)	0.16–0.3 (with sand filtration)	13

pH 6.0, with the addition of ferrous sulfate only rather than at higher pH with lime neutralization. Similar improved treatment, as in the presence of iron, has also been reported for alum flocculation at pH 6.8–7.0 (40), and reflects the phenomena of "coprecipitation," wherein a secondary heavy metal is coprecipitated with a primary metal (iron or aluminum) hydroxide.

Table 26. Treatment of Copper Wastewater (39).

Process	Chemical Addition	pH	Effluent Copper (mg/l) (carbon column)
Iron Only	43 mg/l $FeSO_4$	6.0	0.155
Low Lime	260 mg/l $Ca(OH)_2$ 20 mg/l $FeSO_4$	10.0	0.170
High Lime	600 mg/l $Ca(OH)_2$	11.5	0.352

A significant problem in achieving low residual concentration of copper is the presence of complexing agents, notably cyanide and ammonia. Removal of the complexing agent is essential for good copper treatment. Therefore, copper treatment may depend strongly upon preceding processes. One method of avoiding cyanide treatment is to use the pyrophosphate copper plating process (41). Conventional lime treatment of pyrophosphate is not effective, since the bath is already at pH 8.3–8.8. Elevation of pH to 12.0 with lime will precipitate calcium pyrophosphate, and thus allow simultaneous precipitation of copper hydroxide. High lime treatment has reduced copper pyrophosphate rinse water copper concentration from 33.4 mg/l to less than 1 mg/l (4). The major disadvantages are high lime costs, need for pH readjustment, high volume solids production and low copper content in sludge. The sludge produced in a typical hydroxide precipitation system is acceptable for copper recovery only if the sludge copper concentration is in the range of 6–10% or greater copper, on a dry weight basis. Typical hydroxide sludge copper content has been reported in the range of 0.13–6.0% (dry weight) in sludges of 21–40% total solids (42).

Cuprous oxide, on the other hand, forms a very dense, easy-to-settle, easy-to-handle sludge of 40–60% solids (16,41,42). Treatment of cupric wastewater with hydrazine or some other reducing agent plus sodium hydroxide reduces copper to cuprous form and is recommended by several authors (16,41-43). Such a system may be operated at pH 7–8. A full-scale system has been reported in operation (Table 25), using the integrated treatment method (16). The resulting dense, high-copper sludge is readily marketable. In another application of this technique, copper is reduced from 6–15.5 mg/l to 0.09–0.25 mg/l soluble and 0.30–0.45 total copper. The system is operated with caustic and soda ash at pH 9.5–10.5, and hydrazine at 300–700 mg/l (33).

Limited information is available on sulfide precipitation of copper, principally from mining and ore extraction wastes (19,21). Copper was reduced from 20 mg/l to below detectable limits by sulfide precipitation from a gold-ore extraction wastewater. Treatment by coprecipitation with alum yielded equivalent results (19). In field studies on an acid mine drainage waste, the waste was first lime neutralized to pH 5.0 to precipitate iron and aluminum, and then in a second stage treated at pH 6.5 with barium sulfide for removal of other heavy metals. Copper was reduced from initial levels of 50–115 mg/l, to below 0.5 mg/l (21).

Costs associated with treatment systems involving hydroxide precipitation of copper-bearing wastes are summarized in Table 27. Where possible, costs pertaining directly to copper wastes have been

Table 27. Capital and Operating Costs of Precipitation Treatment of Copper-Bearing Wastewaters.

Process	Flow (MGD)	Capital Costs ($/1000 gpd)	Operating Costs (¢/1000 gal)	Reference
1. Lime Precipitation	0.017	9200 (1957)[a]	354	29
2. Lime Precipitation, Rapid Sand Filtration	2.0	400 (1967)	—	30
3. Lime Precipitation	0.27	1390 (1951)	62	35
4. Integrated, Hydrazine-NaOH (copper recovery)	0.22[b]	653[c]	72[c]	16
5. Lime Precipitation, Chromium Reduction	0.48	461	12.3	44
6. Integrated, Lime (no recovery)	0.24	123	9	44
7. Integrated, Lime, CN Destruction	0.48	96	12.7	44
8. Continuous, Lime, Cr Reduction	0.072	694	28	38

[a] Year of cost data in parentheses if not current.

[b] Based on previous discharge flow prior to recycle. Current blowdown is 0.014 MGD.

[c] Includes cost for electrolytic recovery systems and copper recovery values.

isolated. Frequently, copper removal is associated with cyanide destruction, chromium reduction and nonmetal waste treatment procedures, however. A substantial proportion of capital costs may be related to collection of the wastes. In one reported case, 47% of the capital costs for a metal-finishing waste treatment system was attributed to collection lines (34).

Simple lime precipitation of metal hydroxides should cost 25¢/1000 gal without effluent filtration, and 45¢/1000 gal with filtration (45). If cyanide treatment is part of the complete treatment process, costs could range from 45¢–109¢/1000 gal. The cost advantage of the integrated treatment system is evident from Table 27, but it has been suggested that such systems are not economical for plating plants doing odd jobs and producing a combined waste flow of less than 1000 gph (42).

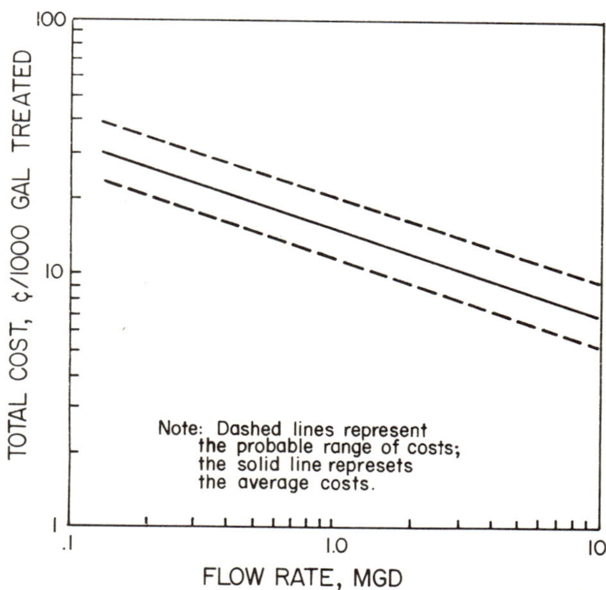

Figure 15. Total cost of treatment by coagulation, sedimentation, and rapid sand filtration. Cost including capital investment (4%, 30 yr.), labor, power, chemicals, maintenance and repair, and heating of buildings (46).

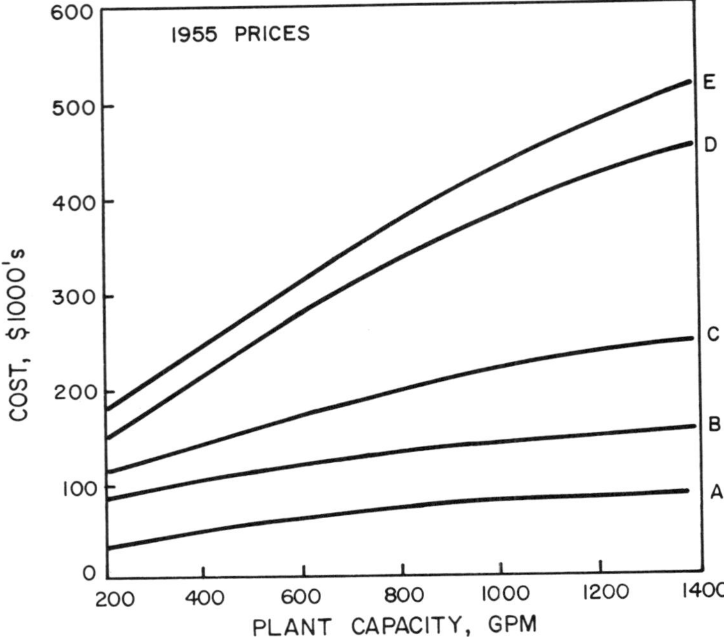

Figure 16. Approximate cost of treatment plant for plating wastes. Curve A—mixing, flocculation and sedimentation equipment only, installed. Curve B—Curve A plus chemical feeders, pumps, and automatic control equipment, installed. Curve C—Curve B plus piping. Curve D—Curve C plus buildings for control, chemical storage, etc. Curve E—Curve D plus engineering, total plant cost. [From Graham (47), courtesy of Van Nostrand Reinhold Publishing Company.]

Evaporative Recovery

Use of evaporative recovery has been practiced for over 20 years (48). Recovery of copper is often more attractive than some other metals such as zinc, due to its higher value. A recent study (49) indicated that the recovery value for copper solutions ranging from 100–500 mg/l is $0.60–$3.00/1000 gal. In brass rinse waters containing 40–250 mg/l of copper, the recovery values are given as $0.20–$1.20/1000 gal. Unless a recovery process is a closed loop operation, however, the concentrated wastewater may still require final treatment for residual metal removal.

Process modification for waste stream segregation and flow reduction is usually an integral part of the recovery process. In fact, direct recovery can result from such modifications. In one case, use of counter-current rinse copper plating baths allowed direct return of the rinse flow into the plating baths as evaporative makeup water (7). Rinse water flow was reduced from 480 to 16 gal/hr for the copper plating lines. The final rinse tank concentration of copper was reduced to 0.9 mg/l, from a previous value of 40 mg/l. In the event that water evaporation from the plating bath is not sufficient to allow direct return of the dilute rinse water, preliminary evaporation of rinse water may be necessary to increase its copper concentration.

Domey and Steifel (25) report that copper plating bath rinse water volume could be reduced from 4100 to 600 gal/day by rinsing modifications. Such modification is generally essential to the economic feasibility of evaporative recovery systems, since the major cost item is the steam (50). Heat requirements may be reduced by using vacuum evaporation, at a somewhat greater capital cost.

One disadvantage of evaporative recovery is the build-up of impurities in the plating bath, since both the impurities added with plating chemicals and those formed by decomposition of additives are returned quantitatively to the plating bath. As these impurities build up, defective plating may result.

Ion Exchange

Ion exchange is capable of achieving very high levels of copper removal, particularly from dilute wastes. Copper removal by ion exchange from 1.02 mg/l to less than 0.03 mg/l has been reported (51). The initial costs of an ion exchange unit depend upon the required resin volume. For dilute wastes, flow rate will be the limiting factor, while the resin exchange capacity will dictate the volume for more concentrated systems. Costs of various units and types of resins are given in Figure 17 as a function of required capacity (15). These costs are for complete units (valves, resin, etc.). Capital investments of $400-$1200/1000 gpd capacity have been cited, with operating costs of 31¢–81¢/1000 gal (44,52,53). When compared to other treatment methods, for cases where metal recovery is not feasible, ion exchange does not appear to be practical from a cost standpoint. Pinner and Crowle (44) were able to make direct comparisons between two nearly identical plating lines producing equal waste volumes (0.48 MGD) of equal composition and concentration. One system employed precipitation treatment and 30% water recycle, the other ion exchange treatment and 90% water recycle.

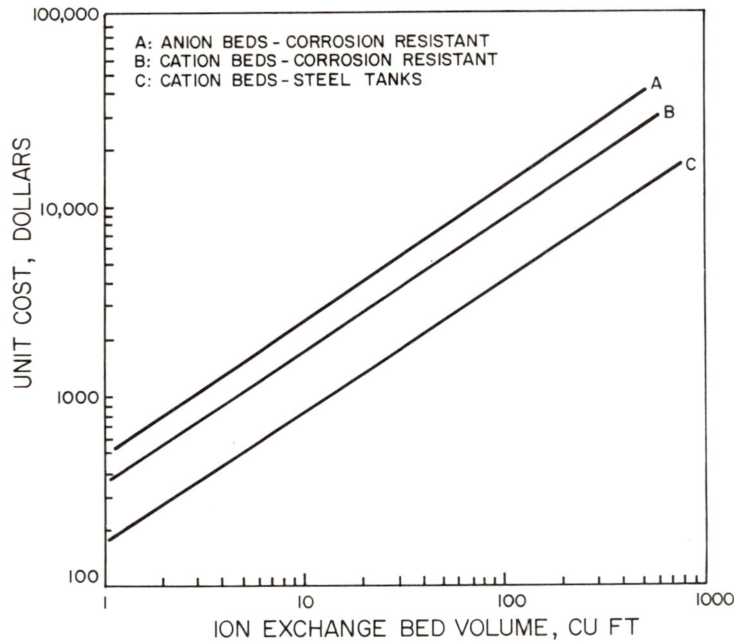

Figure 17. Cost of ion exchange units, complete. [From McGarvey, *et al.* (15), courtesy of *Industrial Engineering Chemistry.*]

Costs were \$461 vs. \$1075/1000 gpd and 12.3¢ vs. 18.2¢/1000 gal treated respectively by precipitation and ion exchange.

Treatment of copper rinse solutions containing high ammonia concentrations has been reported (54). A selective ion exchange resin (Amberlite IR 120, Rohm and Haas) reduced copper from 45 mg/l to undetectable levels. The initial ammonia content of 70 mg/l was reduced to 1 mg/l.

When copper recovery is feasible, ion exchange economics are greatly improved. Schore (55) reported that ion exchange reclamation of copper from pickle liquor allowed electrolytic copper recovery and regenerated the acid pickle solution. He also reported that ion exchange was used to treat 300 gpm (0.432 MGD) of solution from a printed circuit etching machine. In both cases, he claimed that the treatment systems were paid for in 3–4 years by the reclaimed copper.

The major difficulty in copper recovery by ion exchange treatment is to utilize sufficiently small regenerant volume to recover a concentrated copper solution, while adequately regenerating the resin. A counter-current, continuous regeneration system, such as the patented Asahi process (56), is reported to yield regenerant containing 3–4 N copper. In this type of system, of which several are available, the resin moves intermittently down and out of the bed, whereas the wastewater moves up the resin column. Regeneration typically occurs in a tall thin column, to provide efficient resin-regenerant contact. Compared with conventional ion exchange systems, the continuous counter-current system is usually more expensive, complex and inflexible in operation.

Electrolytic Recovery

Direct electrolytic recovery of copper metal is possible from relatively concentrated waste solutions. Treatment of dilute solutions (less than 2 g/l) using conventional electrolysis is impractical, because of the high power consumption (42). The electrolytic process generally requires a preconcentration step such as ion exchange or evaporation. Acid leach solution from copper ore is electrolyzed directly in one application to recover copper. The resultant regenerated sulfuric acid is also recycled (5). Recovery of copper from copper pickling baths presents no problem, but brass pickle solutions may cause problems from tin sludge formation (42). Proper control of electrolysis current can eliminate problems of other metals plating out (16).

For copper cyanide baths, copper recovery by electrolysis can be accomplished in conjunction with cyanide destruction (6,57,58). In the past, the process has not been feasible for dilute solutions but was appropriate for treatment of spent plating bath solutions or ion exchange concentrates. The process economics compare favorably with other copper cyanide treatment processes. Operating data are presented in Table 28, and an economic analysis of the process is given in Table 29.

A new process has been reported recently (24,42,59), which purports to make electrolytic treatment of dilute copper cyanide solutions economically attractive. The process consists of a fluidized semiconducting carbon bed across which a potential is applied. The advantage of this type of system is that it has a very large surface area per unit of volume, which provides a low-current density on the electrode surfaces by a high-current density across the cell. The

Table 28. Reduction of Copper and Cyanide by Electrolysis (57).

Days of Treatment	Copper (g/l)	Cyanide (mg/l)	pH
Start	26	75,000	12.2
1	23	50,000	11.7
4	11	12,500	10.4
6	6	5,980	10.0
8	2	2,200	9.8
11	0	750	9.7
18	—	0.2	9.5

Table 29. Electrolysis Costs per 1000 Gallons of Plating Wastes (57).

Item	Cost ($)
Power consumption	$11.66
Water consumption	.14
Steam consumption	3.50
Plate usage	14.80
Labor	21.00
Total	$51.10
Copper salvage	$24.00
Net Cost	$27.10

system is commercially available in 5- and 10-gpm units (59). It is claimed that 20 mg/l CN is reduced to 0.5 mg/l and 40 mg/l Cu is reduced to below 1.0 mg/l for roughly 5% of conventional treatment costs, at 100 times the rate of conventional electrolysis treating wastes of similar strength (59).

Other Processes

One recently developed process with significant potential application for both concentrated and dilute copper wastewater is cementation (3,20,23,42,60-62). The process involves percolating the copper wastewater through a bed of scrap iron. By oxidation-reduction reaction, copper is cemented onto the iron, with sacrificial iron going into solution. In one pilot scale application, influent copper averaging 79.6 mg/l (range 34–220 mg/l) was reduced to an average effluent of 0.33 mg/l (range 0.2–0.84 mg/l). The copper can be recovered as essentially pure metal, representing a significant offsetting savings to treatment costs.

Bovet (63) has claimed that electrodialysis is economically feasible for treatment of process solutions and rinse waters from plating and metal finishing operations, and for treatment of acid mine drainage. No cost or operating data support this claim. Anderson, *et al.* (45), report operating energy requirements of 10 kwh/1000 gal for electrodialysis and indicate membrane fouling can be a serious problem.

Carbon filtration has been mentioned in conjunction with pilot plant studies on municipal wastes (39,64). Some copper reduction by activated carbon beds was indicated. Pilot plant studies involving the removal of copper cyanide on activated carbon have also been reported (9). For copper cyanide concentrations of 100–113 mg/l copper, copper removals were reported greater than 98% in all cases. Flows were 25–50 gpm.

Successful pilot-scale reverse osmosis treatment (99+% rejection) has been reported for copper pyrophosphate and copper cyanide plating bath rinse waters (65). Advantages claimed for reverse osmosis include low capital and energy costs. However available membranes are suitable only over a pH range of 3.5–12.0 for extended use, and the process is incapable of achieving high solution concentrations. Maximum operating pressure is 400–1500 psig, depending upon membrane type and configuration, and capital costs are reported as $500–$4000/1000 gpd (65).

SUMMARY

Copper treatment is predominantly accomplished by precipitation with lime. Effluent concentrations below 0.5 mg/l are achieved on a consistent basis only with proper pH control and either proper clarifier design accompanied by good sludge settling characteristics or tertiary effluent treatment with sand filtration. Precipitation as the cuprous oxide seems to offer distinct advantages with respect to solids handling, especially when the sludge can be sold for copper reclamation. Final copper concentrations below 0.1 mg/l are reported but do not seem readily attainable on a consistent basis with current precipitation methods. Since copper hydroxide solubility increases at higher pH after passing through a minimum at approximately pH 9–10, pH variations in the waste stream may be a causative factor in limiting the removal of copper.

Process modification and waste stream separation provide a means of enhanced treatment of copper-bearing wastes. When such changes are implemented, recovery processes become more attractive. In fact, to be economically feasible ion exchange treatment would seem to require recovery, unless available space is the limiting factor. Recovery of one type or another is practiced in many industrial operations, and appears economically attractive, primarily due to the high value of recovered copper.

REFERENCES

1. Golomb, A. "Application of Reverse Osmosis to Electroplating Waste Water Treatment. Part II. The Potential of Reverse Osmosis in the Treatment of Some Plating Wastes," *Plating* **59**, 316-319 (1972).
2. Rudolfs, W. (Ed.), *Industrial Waste Treatment,* (New York: Van Nostrand Reinhold Co., 1953).
3. Dean, J. G., F. L. Bosqui and K. H. Lanouette. "Removing Heavy Metals from Waste Water," *Environ. Sci. Technol.* **6**, 518-522 (1972).
4. Parsons, W. A. and W. Rudolfs. "Lime Treatment of Copper Pyrophosphate Plating Wastes," *Sewage Ind. Wastes Eng.* **22**, 313-315 (1951).
5. Simpson, R. W. and K. Thompson. "Chlorine Treatment of Cyanide Wastes," *Sewage Ind. Wastes Eng.* **21**, 302-304 (1950).
6. Pinkerton, H. L. "Waste Disposal," In *Electroplating Engineering Handbook,* 2nd edition, A. Kenneth Graham, Editor-in-chief, (New York: Van Nostrand Reinhold Co., 1962).
7. Barnes, G. E. "Disposal and Recovery of Electroplating Wastes," *J. Water Poll. Cont. Fed.* **40**, 1459-1470 (1968).
8. Wise, W. S. "The Industrial Waste Problem. IV. Brass and Copper Electroplating and Textile Wastes," *Sewage Ind. Wastes* **20**, 96-102 (1948).
9. Battelle Columbus Laboratories. "An Investigation of Techniques for Removal of Cyanide from Electroplating Wastes," U.S. EPA Report 12010 EIE 11/71 (1971).
10. Lowe, W. "The Origin and Characteristics of Toxic Wastes with Particular Reference to the Metals Industries," *Water Poll. Cont. (London)* 270-280 (1970).
11. Anderson, J. S. and E. H. Iobst, Jr. "Case History of Wastewater Treatment in a General Electric Appliance Plant," *J. Water Poll. Cont. Fed.* **40**, 1786-1795 (1968).
12. Gard, C. M., C. A. Snavely and D. J. Lemon. "Design and Operation of a Metal Wastes Treatment Plant," *Sewage Ind. Wastes* **23**, 1429-1438 (1951).
13. Nemerow, N. L. *Theories and Practices of Industrial Waste Treatment,* (Reading, Massachusetts: Addison-Wesley Publishing Co., 1963).
14. McGarvey, F. X. "The Application of Ion Exchange Resins to Metallurgical Waste Problems," *Proc. 7th Purdue Industrial Waste Conf.* **7**, 289-304 (1952).
15. McGarvey, F. X., R. E. Tenhoor and R. P. Nevers. "Brass and Copper Industry: Cation Exchangers for Metal Concentration from Pickle Rinse Waters," *Ind. Eng. Chem.* **44**, 534-541 (1952).
16. Volco Brass and Copper Co. "Brass Wire Mill Process Changes and

Waste Abatement, Recovery and Reuse," U.S. EPA Report 12010 DPF 11/71 (1971).

17. Tallmange, J. A. "Nonferrous Metals," In *Chemical Technology, Vol. 2, Industrial Waste Water Control,* C. Fred Gurham, Ed. (New York: Academic Press, Inc., 1965).
18. Hallowell, J. B., J. F. Shea, G. R. Smithson, Jr., A. B. Tripler and B. W. Gonser. "Water Pollution Control in the Primary Non-Ferrous Metals Industry, Volume I. Copper, Zinc, and Lead Industries," U.S. Environmental Protection Agency, Report EPA-R2-73-247a (September, 1973).
19. Rosehart, R. and J. Lee. "Effective Methods of Arsenic Removal from Gold Mine Wastes," *Can. Mining J.* 53-57 (June, 1972).
20. Hill, R. D. "Control and Prevention of Mine Drainage," Presented at the Environmental Resources Conference on Cycling and Control of Metals, Battelle Columbus Laboratories, 1972.
21. Larsen, H. P., J. K. P. Shou and L. W. Ross. "Chemical Treatment of Metal-Bearing Mine Drainage," *J. Water Poll. Cont. Fed.* **45** (8), 1682-1695 (1973).
22. Chalmers, R. K. "Pretreatment of Toxic Wastes," *Water Poll. Cont. (London)* 281-291 (1970).
23. Ciancia, J. "Pollution Abatement in the Metal Finishing Industry," Presented at the Environmental Resources Conference on Cycling and Control of Metals, Battelle Columbus Laboratories, 1972.
24. Silman, H. "Treatment of Rinse Water from Electrochemical Processes," *Metal Finishing* **69**, 62-66 (1971).
25. Domey, W. R. and R. C. Stiefe. "Wastewater Reduction in Metal Finishing Operations," Presented at 43rd Annual Conf. Water Poll. Cont. Fed., 1970.
26. Foulk, D. G. "Metal Finishing Products," In *Chemical Technology, Vol. 2, Industrial Waste Water Control,* C. Fred Gurnham, Ed. (New York: Academic Press, Inc., 1965).
27. Stumm, W. and J. J. Morgan. *Aquatic Chemistry,* (New York: Wiley-Interscience, 1970).
28. Jenkins, S. H., D. G. Knight and R. E. Humphreys. "The Solubility of Heavy Metal Hydroxides in Water, Sewage and Sewage Sludge. I. Solubility of Some Metal Hydroxides," *Internat. J. Air Water Poll.* **8**, 537-556.
29. Nyquist, O. W. and H. R. Carroll. "Design and Treatment of Metal Processing Wastewaters," *Sewage Ind. Wastes* **31**, 941-948 (1959).
30. Stone, E. H. F. "Treatment of Non-ferrous Metal Process Waste of Kynoch Works, Birmingham, England," *Proc. 25th Purdue Ind. Waste Conf.* **25**, 848-865 (1967).
31. Stone, E. H. F. "Treatment of Non-Ferrous Metal Process Waste," *Metal Finishing J.* **18** (212), 280-290 (1972).

32. Ogawa, H., Y. Makoto and K. Hurikawa. "Industrial Waste-Water Treatment. Treatment of Waste Water Containing Copper Ions," *Mizu Shari Gijustu (Japan)* **13** (2), 15-20 (1972).
33. Martin, J. J., Jr. "Chemical Treatment of Plating Waste for Removal of Heavy Metals," U.S. EPA Report EPA-R2-73-044 (May, 1973).
34. Hansen, N. H. and W. Zabban. "Design and Operation Problems of a Continuous Automatic Plating Waste Treatment Plant at Data Processing Division, IBM, Rochester, Minnesota," *Proc. 14th Purdue Ind. Waste Conf.* **14,** 227-249 (1959).
35. Watson, K. S. "Treatment of Complex Metal-Finishing Wastes," *Sewage Ind. Wastes,* **26,** 182-194 (1954).
36. "Effluent Treatment at BSR," *Metal Finishing J.* **17,** 248 (1971).
37. Chalmers, R. K. "Trade Effluent Treatment at the Cannock Factory of Joseph Lucas, Ltd.," *J. Proc. Inst. Sewage Purif.,* 357-359 (1965).
38. Teer, E. H. and L. V. Russell. "Heavy Metals from Wood Preserving Wastewater," Presented at the 27th Ind. Waste Conf., Purdue University, 1972.
39. Maruyama, T., S. A. Hannah and J. M. Cohen. "Removal of Heavy Metals by Physical and Chemical Treatment Processes," Presented at the 45th Annual Water Poll. Cont. Fed. meeting, 1972.
40. Culp, G. L. and R. L. Culp. *New Concepts in Water Purification,* (New York: Van Nostrand Reinhold Co., 1974).
41. Crowle, V. "Effluent Problems as They Effect the Zinc Die-Casting and Plating on Plastics Industries," *Metal Finishing J.* **17,** 51-54 (1971).
42. Jackson, D. V. "Metal Recovery from Effluents and Sludges," *Metal Finishing J.* **18,** 235-242 (1972).
43. Curry, N. "Philosophy and Methodology of Metallic Waste Treatment," Presented at the 27th Ind. Waste Conf., Purdue University, 1972.
44. Pinner, R. and V. Crowle. "Cost Factors for Effluent Treatment and Recovery of Materials in the Metal Finishing Department," *Electroplating Metal Finishing* **3,** 13-31 (1971).
45. Anderson, J. S., J. M. Philips and C. B. Schriver. "Water Contamina-ation by Certain Heavy Metals," Presented at the 44th Annual Water Poll. Cont. Fed. meeting, 1971.
46. "Cost of Wastewater Treatment Processes," Robert A. Taft Research Center, Report TWRC-6, (Washington, D.C.: U.S. Dept. of the Interior, 1968).
47. Graham, A. K. *Electroplating Engineering Handbook,* (New York: Van Nostrand Reinhold Co., 1962).
48. Culotta, J. M. and W. F. Swanton. "Case Histories of Plating Waste Recovery Systems," *Plating* **57,** 251-255 (1970).

49. *State of the Art; Review on Product Recovery,* Water Pollution Control Research Series, 1707 ODJW 11/69, (Washington, D.C.: U.S. Dept. of the Interior, 1969).
50. Culotta, J. M. and W. F. Swanton. "Controls for Plating Recovery System," *Plating* **58,** 783-785 (1971).
51. Botham, G. H. and W. R. Bryson. "Note on the Corrosion of Aluminum Dairy Equipment by Water," *J. Dairy Res.* **20,** 154-155 (1953).
52. Lancy, L., W. Nohse and D. Wystroch. "Practical and Economic Comparison of the Most Common Metal Finishing Waste Treatment Systems," *Plating* **59,** 126-130 (1972).
53. Sievers, J. F. and C. J. Novotny. "Recovery of Mixed Rinse Water by Means of Ion Exchange," *Plating* **42,** 482-485 (1971).
54. "Winning Heavy Metals from Waste Streams," *Chem. Eng.* **78,** 62-64 (1971).
55. Schore, G. "Electronic Equipment and Ion Exchange for Use in Automated Treatment Systems," Presented at 27th Ind. Waste Conf., Purdue University, 1972.
56. Solt, G. S. "Waste Treatment by Ion Exchange," *Effluent Water Treat. J.* **13** (12), 768-773 (1973).
57. Easton, J. K. "Electrolytic Decomposition of Concentrated Cyanide Plating Wastes," *J. Water Poll. Cont. Fed.* **39,** 1621-1625 (1967).
58. American Electroplating Society. "A Report on the Control of Cyanides in Plating Shop Effluents," *Plating* **56,** 1107-1112 (1969).
59. "Electrolysis Speeds Up Waste Treatment," *Environ. Sci. Technol.* **4,** 201 (1970).
60. Jester, T. L. and T. H. Taylor. "Industrial Waste Treatment at Scoville Manufacturing Company," Presented at 28th Purdue Ind. Waste Conf., 1973.
61. Case, O. P. "Metallic Recovery from Waste Waters Utilizing Cementation," U.S. EPA Report No. EPA-670/2-74-008 (January, 1974).
62. Anderson, J. R. and R. P. Hollingworth. "Removal of Copper from Aqueous Solutions with Highly Stressed Activated Ferrous Metals," U.S. Patent 3,674,446 (July 4, 1972).
63. Bovet, E. D. "Industrial Desalting for By-Product Recovery," *J. Amer. Water Works Assoc.* **62,** 539-542 (1970).
64. Nilsson, R. "Removal of Metals by Chemical Treatment of Municipal Waste Water," *Water Res.* **5,** 51-60 (1971).
65. Donnelly, R. G., R. L. Goldsmith, K. J. McNulty and M. Tan. "Reverse Osmosis Treatment of Electroplating Wastes," *Plating* **61** (5), 432-442 (1974).

9

TREATMENT TECHNOLOGY FOR CYANIDE

INDUSTRIAL SOURCES

Cyanide, as sodium cyanide (NaCN) or hydrocyanic acid (HCN) is a widely used industrial material. Cyanide waste streams result from ore extracting and mining, photographic processing, coke furnaces, synthetic fiber manufacturing, case hardening and pickling of steel, and industrial gas scrubbing. A major source of waste cyanide is the electroplating industry. Electroplaters use cyanide baths to hold metallic ions such as zinc and cadmium in solution. "Dragover" of the plating solution, containing cyanide ions and metal-cyanide complexes, contaminates rinsing baths. Table 30 presents typical cyanide levels reported for plating wastewaters.

TREATMENT TECHNOLOGY

Several methods of treating cyanide wastes are in current use. The process most frequently employed is cyanide destruction by chlorination. Cyanide may be either partially oxidized to cyanate (CNO^-), or completely oxidized to carbon dioxide (CO_2) and nitro-

Table 30. Concentrations of Cyanide in Plating Wastewaters.

Process	Average (mg/l)	Range (mg/l)	Reference
Plating Rinse	2	0.3–4	1
Plating Rinse	700		2
Plating Rinse		10–25	3
Plating Rinse	32.5		4
Plating Rinse	25		5
Plating Rinse		60–80	6
Plating Rinse		30–50	7
Plating Rinse	55.6	1.4–256	1
Bright Dip		15–20	7
General (Separate Cyanide)	72	9–115	8
General (Combined Stream)	28	1–103	8
Alkaline Cleaning Bath		4,000–8,000	7
Plating Bath	30,000		9
Plating Bath		45,000–100,000	10
Plating Bath			7
Brass		16,000–48,000	
Bronze		40,000–50,000	
Cadmium		20,000–67,000	
Copper		15,000–52,000	
Silver		12,000–60,000	
Tin-Zinc		40,000–50,000	
Zinc		4,000–64,000	

gen (N_2). Another cyanide destruction process, electrolytic decomposition, is applicable where waste cyanide concentrations are very high. Ozone oxidation of cyanide has been employed with some success, and is claimed to be equivalent to destruction by chlorination, on the basis of effectiveness and costs. Evaporative recovery has also been successfully employed to treat cyanide wastewaters. Other cyanide treatment processes, which are primarily in the development stage, include reverse osmosis, ion exchange, and catalytic oxidation.

Chlorination

Destruction of cyanide by chlorination may be accomplished by direct addition of sodium hypochlorite, or by addition of chlorine gas plus sodium hydroxide to the waste. Sodium hydroxide reacts with the chlorine to form sodium hypochlorite. Selection between the two methods is on the basis of economics and safety. Chlorine gas treatment is about half as expensive as direct hypochlorite addition, but handling is more dangerous and equipment costs are higher. The hypochlorite added or produced oxidizes cyanide to cyanate. Oxidation of cyanide to cyanate is accomplished most completely and rapidly under alkaline conditions at pH 10 or higher, and this cyanide treatment process is often termed alkaline chlorination. An oxidation period of 30 minutes to 2 hours is usually allowed (6). To avoid solid cyanide precipitates which may resist chlorination, the wastewater should be continuously mixed during treatment.

The resulting cyanate is much less toxic than cyanide (11), but can if necessary be further oxidized to carbon dioxide and nitrogen. Oxidation of cyanate to carbon dioxide and nitrogen can be accomplished by additional chlorination, or to carbon dioxide and ammonia by acid hydrolysis. Acid hydrolysis must take place at pH 2–3, however, usually obtained by addition of sulfuric acid. Thus, acid hydrolysis requires addition of acid to achieve the required pH, and subsequent neutralization of the acidic wastewater before the effluent can be discharged. These processes of pH adjustment increase the total dissolved solids of the wastewater. Cyanate will be destroyed by acid hydrolysis within about five minutes' contact time.

Complete cyanide oxidation to CO_2 and N_2 can be accomplished solely through chlorination, providing close pH control is maintained. After initial oxidation to cyanate, further oxidation to yield CO_2 and N_2 will occur slowly over several hours at pH 10+. However, if the waste pH is reduced to 8–8.5, cyanate oxidation by chlorine can be completed within one hour. At the lower pH, sufficient chlorine must be added to insure an excess beyond that needed to oxidize the cyanide to cyanate, to avoid liberation of highly toxic cyanogen chloride. Cyanogen chloride is the intermediate product of the oxidation of cyanide to cyanate. It breaks down very rapidly at pH 10+ and temperatures above 20°C (12). However, at lower pH or temperature, excess chlorine is needed to speed the breakdown (11). Complete cyanide treatment at high pH, although requiring long reaction times, may be advantageous if the treated wastewater can be utilized to neutralize acidic waste streams in the plant.

Cyanide baths and rinses are frequently treated together, with the concentrated bath solutions being blended into the more dilute rinse wastewaters. It is common practice to use batch treatment for alkaline chlorination when average flows do not exceed 15 gpm. Batch processes have been used for waste flows up to 30 gpm, however (13).

One modification of the batch process is the patented integrated process (14). Used in plating shops, when articles leave any solution which contains cyanide, they first enter a rinse which consists of a concentrated solution of sodium hypochlorite. Drag-out cyanide is rapidly oxidized to cyanate at pH 10–12.5 and the free chlorine content of 0.3–1.0 g/l of this static rinse bath. The cyanate slowly decomposes to carbon dioxide and nitrogen. It is claimed that any metal in the dragout, which was previously complexed by the cyanide is released by the oxidation of the cyanide and, at the pH maintained is precipitated (14). After the work piece leaves the static treatment bath, it is rinsed with water.

Several manufacturers have developed small "package" cyanide treatment plants, which employ the chlorination process. Figure 18 shows basic types of package plants (15). The tank-type plant shown in Figure 18(a) provides cyanide treatment to the level of cyanate only. If complete destruction is required, a second tank is employed for secondary pH control and second-stage chlorination. The two-stage reaction tower-type plant shown in Figure 18 (6) provides complete cyanide treatment to the level of CO_2 and N_2. Package plants such as these range from treatment capacities of 800 gallons of waste per hour containing a maximum of 450 mg/l of cyanide, to units of 1800 gal/hr and a maximum cyanide content of 350 mg/l. A unit of 800 gal/hr capacity costs about $15,000 complete (16). Investments of $14,250 to $18,000 are quoted for automated continuous package plants designed for complete (two-stage) cyanide oxidation (17).

O'Neill has described a 1000 gal/hr two-stage cyanide oxidation plant (9). Chlorine gas and caustic are used to treat a cyanide cleaning bath rinsewater to less than 0.4 mg/l effluent cyanide. The treatment plant, including a chromate reduction and precipitation process, cost $16,000 for equipment plus another $10,000 for the building and piping. The cyanide treatment segment was continuous, but not automated.

Larger volumes of cyanide waste require especially designed and constructed treatment plants, tailored to the individual industrial process and flow configuration producing the waste. Fisco (18) has described such a treatment plant, which oxidizes the cyanide component of a waste stream totaling 1.5 mgd. Treatment involves total cyanide destruction by a two-step process of chlorination at high and low alkaline pH as discussed above.

Treatment of a cyanide waste flow of 2500–3000 gal/hr by alkaline chlorination to the level of cyanate has been reported (19). An effluent cyanide content of 0.1 mg/l was achieved by the process. Complete oxidation of cyanide to CO_2 was achieved in treating

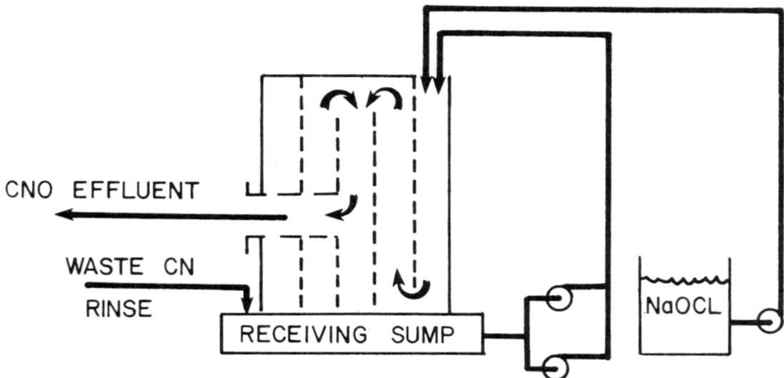

a) Retention tank-type package cyanide treatment plant.

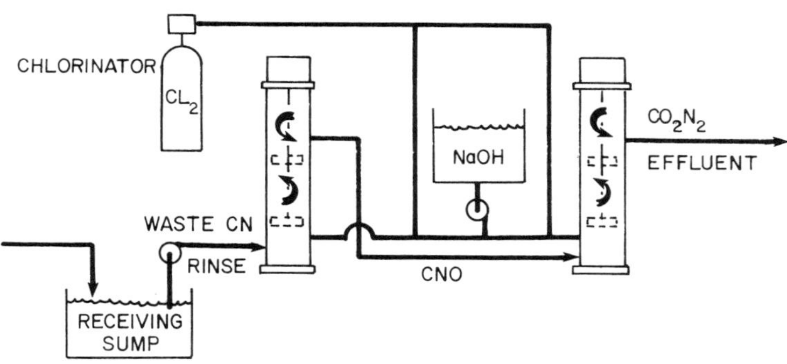

b) Reaction tower-type package cyanide treatment plant.

Figure 18. Commercial package plants for cyanide waste treatment. [From Zievers, *et al.* (15), courtesy of *Plating.*]

another waste by alkaline chlorination plus acid hydrolysis. Cyanide content of the rinse water was reduced from 700 mg/l to zero cyanide plus cyanate (2).

The cyanide content of electroplating rinse waters are quite variable (Table 30). Watson (4) has reported the treatment of a dilute cyanide waste, for a metal plating plant. The waste contained an average cyanide content of 32.5 mg/l, and was treated to zero cyanide, on a batch basis at the rate of 200,000 gallons per month. Effluent cyanate concentration was not reported. Cyanide treatment was to the level of cyanate by alkaline chlorination; three pounds of chlorine being fed per pound of cyanide treated.

It has been reported (11) that oxidation of cyanide to cyanate requires approximately 1.75 pounds of chlorine per pound of cyanide, and complete oxidation to CO_2 plus N_2 requires about 4.3 pounds of chlorine per pound of cyanide. Others, however, disagree with this dosage, reporting that 3.4–4.0, and 8 pounds of chlorine per pound of cyanide are respectively required for the two levels of oxidation (13,20).

Complete destruction of cyanide to CO_2 would increment the treatment costs above those presented above for simple oxidation to cyanate. Cost curves for single-stage cyanide treatment (to cyanate), and for complete cyanide oxidation have been published. Costs, which are reproduced in Figures 19 and 20, indicate single-stage treatment costs of $0.28–$1.28/lb cyanide destroyed, and two-stage costs of $0.50–$2.45/lb cyanide. Capital costs are also shown in both figures, expressed as thousands of dollars capital investment versus waste flow in gallons per minute.

Unless metal-bearing wastes are segregated from the cyanide wastes, an extra chlorine demand will be exerted as a result of metal oxidation by the chlorine (*e.g.*, Ni^{+2} to Ni^{+3}, Fe^{+2} to Fe^{+3}, Cr^{+3} to Cr^{+6}, etc.) Any extra chlorine demand exerted by a mixed metal-cyanide waste would increase cyanide treatment costs significantly.

One major problem associated with alkaline chlorination of cyanide wastes occurs if soluble iron is present with the cyanide. Iron, in the presence of cyanide, forms extremely stable ferrocyanide complexes, which prevent the cyanide from being oxidized (21). Similar difficulties result from formation of nickelocyanides (22). Successful treatment of cyanide in ferri- and ferrocyanide wastes has been reported however by alkaline chlorination at pH 10+ and a temperature 90°C (23). Cyanide was reduced from 5.1 to 0.1 mg/l. Curry (24) in discussing treatment of ferrocyanides, suggests precipitation with ferrous salts as the ferro-ferrocyanide, or alkaline chlorination at a temperature of 71°C and pH of 12.0. The ferro-ferrocyanide precipitation process does not appear to be very effective, as effluent cyanide levels of 5–10 mg/l have been reported (25).

Electrolytic Decomposition

Wastes containing high concentration of cyanide are most successfully treated by electrolytic decomposition. Electrolytic decomposition is a well-established technique, primarily used by industry for the destruction of cyanide in concentrated spent stripping and dip solutions for copper and nickel, alkaline descalers and derusters. It

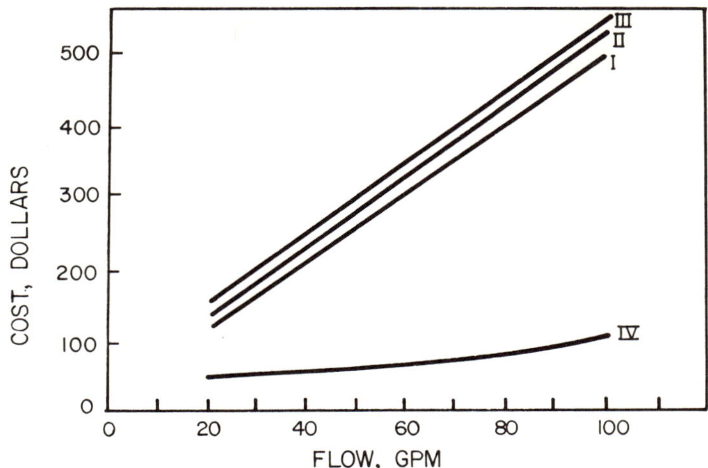

a) Monthly operating costs (plus depreciation) at 50 mg/l of cyanide, 172 hr/month. I—no instrumentation; II—medium instrumentation; III—full instrumentation; IV—using "low-cyanide" type bath.

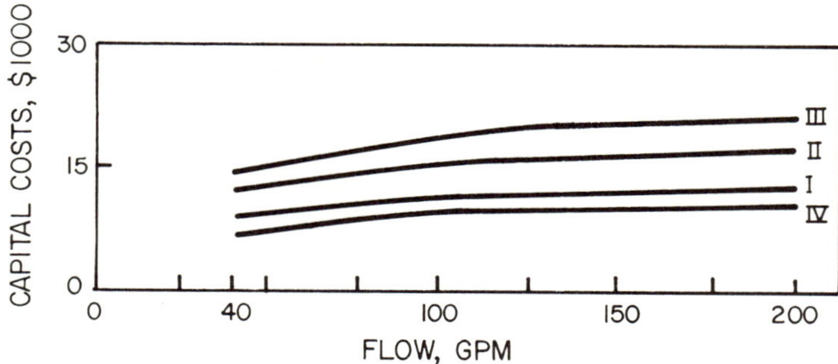

b) Initial capital costs, including installation. I—manual controls; II—medium instrumentation; III—full instrumentation; IV—package plant.

Figure 19. Costs of single cyanide treatment to the level of cyanate. [From Zievers, *et al.* (15), courtesy of *Plating*.]

has not generally been practical to treat rinse waters by this method, due to the initial low cyanide concentrations (3).

The concentrated cyanide waste is subjected to anodic electrolysis at high temperature (125–200°F) for several days. Initially,

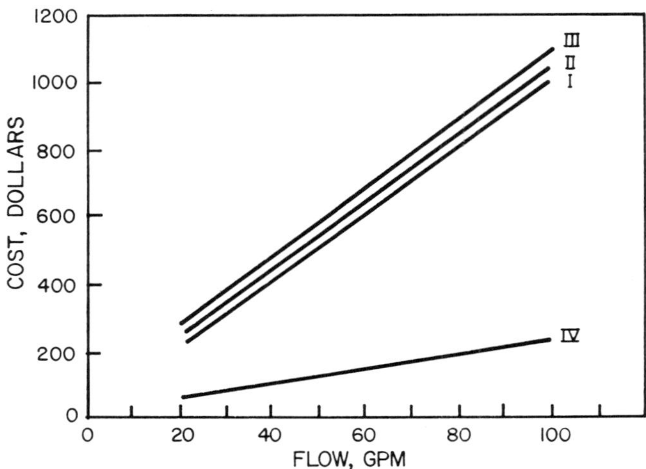

a) Monthly operating costs (plus depreciation) at 50 mg/l cyanide, 172 hr/month. I—no instrumentation; II—medium instrumentation; III—full instrumentation; IV—using "low cyanide" type bath.

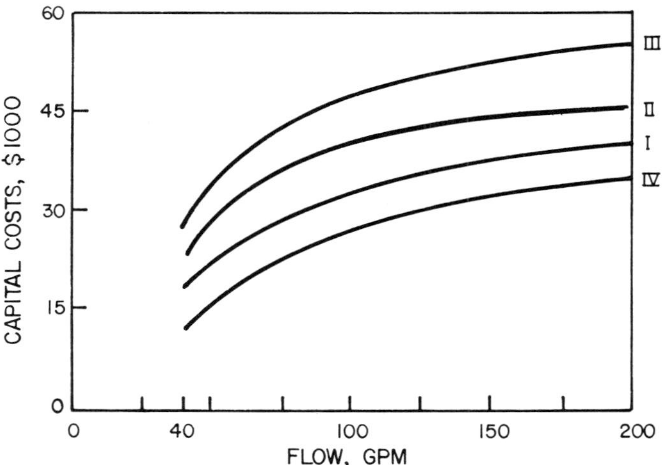

b) Initial capital costs, including installation. I—manual controls; II—medium instrumentation; III—full instrumentation; IV—package plant.

Figure 20. Costs of complete cyanide destruction to the level of carbon dioxide. [From Zievers, *et al.* (15), courtesy of *Plating*.]

cyanide is completely broken down to the gaseous products carbon dioxide, nitrogen, and ammonia, with cyanate as an intermediate (20). As the process continues, the waste-electrolyte becomes less capable of conducting electricity, and the reaction may not go to final completion. Some residual cyanate may be formed, which requires chlorination to complete the waste treatment. Low levels of residual cyanide can be achieved by electrolytic decomposition, providing sufficient treatment time is allowed (10). Results of treatment of several batches of plating wastes are presented in Table 31. Wastes containing several tens of thousands of mg/l of cyanide were treated, over 7 to 18-day periods, down to cyanide levels of less than 0.5 mg/l. Electrolytic decomposition is less effective on cyanide wastes containing sulfate. Cyanide reductions to only 695–750 mg/l have been reported, before heavy scaling at the anode prevented further electrolysis (10). Initial cyanide concentrations were equivalent to those presented in Table 31. Sulfate concentrations in the cyanide wastes were not reported.

Table 31. Electrolytic Decomposition of Cyanide Waste (10).

Run No.	Initial Cyanide Conc. (mg/l)	Time to Decompose (days)	Final Cyanide Conc. (mg/l)
1	95,000	16	0.1
2	75,000	17	0.2
3	50,000	10	0.4
4	75,000	18	0.2
5	65,000	12	0.2
6	100,000	17	0.3
7	55,000	14	0.4
8	45,000	7	0.1
9	50,000	14	0.1
10	55,000	8	0.2
11	48,000	12	0.4

Operating costs reported for electrolytic decomposition were 8.2¢/lb of cyanide destroyed (10). At cyanide waste concentrations shown in Table 31, these costs convert to $55.00/1000 gal.

If sodium chloride is added to the cyanide solution the oxidation rate, power consumption and efficiency of electrolytic decomposition at lower cyanide concentrations (below 200 mg/l) is greatly enhanced (14,20). In addition to improving solution conductivity, the anodic oxidation of the chloride ion to chlorine gas yields chlorination oxidation concurrent with electrolytic destruction. One particular advantage for electrolytic decomposition is its effectiveness in destroying nickel, copper, ferro and ferricyanide complexes, which are difficult to treat by alkaline chlorination (20).

A package electrolytic decomposition treatment plant has been developed specifically to treat rinse waters containing low cyanide content, and is claimed to purify a waste stream with 20 mg/l to less than 0.5 mg/l of cyanide (26). Unit capacity is 5–10 gpm and capital cost is less than $20,000 for the typical plating shop (26).

Ozonation

Ozonation shows some promise as a substitute for chlorination of cyanide waste, although it has had limited full-scale use. Besselievre (10) has pointed out that ozonation offers perhaps the cheapest method of destroying cyanide, all things considered, because the cost of producing ozone, excluding capital costs, is low and the amount of ozone required per pound of cyanide is small. Ozone oxidation of cyanide to cyanate requires 1.8–2 lb/lb CN, and complete oxidation 4.6–5 lb/lb CN. An additional advantage is that no dissolved solids are added in the treatment step. Kandzas and Mokina (27) have reported complete ozone oxidation of cyanide (via cyanate intermediate) to end products. Complex zinc, nickel, and copper cyanides were easily destroyed, but complex cobalt cyanide was resistant to treatment by ozonation.

Ozonation has been employed to treat a cyanide waste at a metal working plant (5). Waste flow through the plant is 500 gpm, and cyanide concentrations to 25 mg/l were encountered. Oxidation of cyanide to cyanate was reported, with partial oxidation of cyanate to final end products. Capital costs for the process, installed, totalled $70,000. Ozone production costs of 14–15¢/lb (20) compared with chlorine costs of 13¢/lb (15).

In ozone treatment, cyanide oxidation to cyanate is very rapid (10–15 min) at pH 9–12, and is practically instantaneous in the presence of traces of copper, which acts as a catalyst (20). Oxidation of cyanate to final end products is much slower, especially in the case of complex cyanides other than copper and nickel, unless catalysts such as iron, copper or manganese are present.

Other Processes

Recently, a patented process (Kastone, E. I. du Pont de Nemours and Co.) has become commercially available. The process utilizes hydrogen peroxide plus formaldehyde to treat cyanide wastes at

120–130°F. End products include cyanate, ammonia, and various organic acids (28). The process simultaneously precipitates zinc and cadmium, if present. At one commercial installation, the Kastone process reduced cyanide from initial values of 350–400 mg/l to less than 0.3 mg/l (29). The process is reported effective for cyanide levels of 50–500 mg/l. Treatment costs of $0.41/lb cyanide were cited, which were compared to alkaline chlorination costs of $0.18/lb (28). Additional economic benefits accrue to the Kastone process, due to the simultaneous precipitation of zinc or cadmium. One major disadvantage of the process is its inability to oxidize cyanide below the level of cyanate.

Evaporative recovery of cyanide wastes has been shown to be practical and economical under proper circumstances. Its use is primarily for concentrated rinse waters. The evaporation process is best applied for segregated plating lines with counter-current rinse tanks. Under these conditions, the system may operate as a closed loop, with recondensed water vapor returned to the head of the rinse line, and concentrated plating solution returned to the plating tank. A schematic of a typical closed loop evaporative recovery system is shown in Figure 21.

Evaporative recovery is perhaps best suited to treatment of cya-

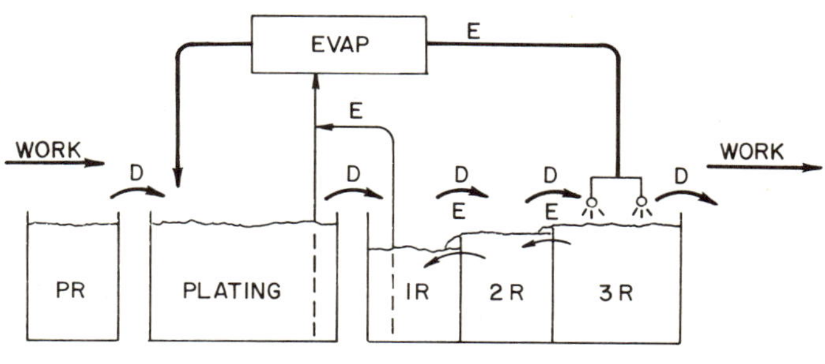

PR = PRERINSE TANK
D = DRAGOUT
R = RINSE TANK
P = RETURNED PLATING
E = RINSE OR DISTILLATE FLOW

Figure 21. Schematic of a closed loop evaporative recovery system. [From Culotta and Swanton (30), courtesy of *Plating*.]

nide plating solution above all others, due to the non-corrosive nature of the distillate. This results in lower equipment and maintenance costs. Operating costs are primarily influenced by cost of steam, cooling water and power. For fixed costs of steam, water and power, utilities costs become a direct function of evaporative rate. Counter-current rinsing, wherein only the most contaminated rinse water is treated, thus improves the economy of the process. Reduced utilities cost of multiple effect evaporation results from improved efficiency of steam usage. This is offset by increased capital cost for each additional evaporation effect (31).

Other cyanide treatment processes include ion exchange (32), reverse osmosis (33,34), and copper-catalyzed activated carbon adsorption and oxidation (1,25). Ion exchange recovery of cyanide does not yet appear feasible because, among other reasons, cyanide deteriorates the exchange resins (13). Metal cyanide fouling has also been reported (35). Zievers and Novotny (32) report ion exchange effluents containing 6–15 mg/l of cyanide, and point out the danger of evolution of highly toxic hydrogen cyanide gas upon resin regeneration. One potential application of ion exchange is the use of weak base anion exchange resin to remove ferri- and ferrocyanide (12). Beyond this limited scope however, ion exchange is not generally considered feasible for cyanide wastewaters.

Cellulose acetate and polymide reverse osmosis membranes deteriorate rapidly by alkaline hydrolysis at the normal pH of cyanide wastes (pH 11+). Reduction of pH precipitates metals in the waste, and clogs the membrane. More resistant membranes have shown poor performance, rejecting only 28% of the cyanide (33). Further, most effective membrane rejection of cyanide occurs at pH above 10 (36). Field applications of reverse osmosis do not appear likely, pending development of membranes possessing a higher rejection capacity at moderate pH, or greater resistance to hydrolysis at alkaline pH.

A recently developed process involving cyanide adsorption plus copper catalyzed oxidation on granular activated carbon has proven effective in preliminary pilot plant tests (1,20,37). Under controlled conditions the process yielded effluent cyanide levels below 0.1 mg/l for a wide range of influent cyanide concentration. The process has not yet been tested on a full-scale long-term basis, however.

SUMMARY

Processes are available to treat cyanide wastewaters to almost any effluent level required. Treatment of strong cyanide wastes appears to be most feasible by electrolytic decomposition. Low to intermediate effluent cyanide content is reported by this process, and either partial treatment to cyanate or complete destruction can be achieved as required. Chemical oxidation by chlorination or ozonation is appropriate for relatively dilute cyanide wastes, whether carried over from the electrolytic decomposition process or as dilute rinse water wastes. If proper rinsing procedures are used, yielding a concentrated cyanide rinse water, evaporative recovery is both feasible and economical. This results in zero cyanide discharge, plus reuse of distillate and the concentrated cyanide solution. Table 32 presents a summary of cyanide treatment processes and reported efficiencies. For the alkaline chlorination process, improved treatment efficiencies result from proper pH control and increased chlorine dosages.

Table 32. Treatment Levels for Cyanide Wastewaters.

Treatment Process	Cyanide Concentration (mg/l) Initial	Final	Percent Removal	Reference
Alkaline Chlorination (1)	—	1.7	—	38
Alkaline Chlorination (1)	—	0.1	—	19
Alkaline Chlorination (2)	—	0.4	—	9
Alkaline Chlorination (2)	700	0.0	100	2
Alkaline Chlorination	32.5	0.0	100	4
Alkaline Chlorination (3)	5.1	0.1	98	23
Ferro-Ferrocyanide Precipitation	—	5–10	—	25
Electrolysis	45,000–100,000	0.1–0.5	99.99+	10
Electrolysis	20	0.5	97.5	26
Ozonation	25	0	100	5
Kastone Process	350–400	0.3	99.9+	29

Notes: (1) Single-Stage Chlorination
(2) Two-Stage Chlorination
(3) High pH-high temperature process

Capital investments for alkaline chlorination plants, as reported by several authors (9,15-17), appear to be accurately described by the costs of Figures 19 and 20. Green and Smith (20) have reported investments for common cyanide treatment processes by English industries. Their data, converted to cost ratios, are presented in Table 33. The least expensive alkaline chlorination plant has been taken as a base cost of one unit, and all other capital investments compared to that. Increased capital investments are often offset by lower operating costs, however. This is shown in Table 34, which summarizes reported operating costs.

Table 33. Comparison of Capital Costs for Cyanide Treatment Process (After 20).

Treatment Process	Plant Capacity (1000 gpd)	Cost Ratio (per 1000 gpd)
Alkaline Chlorination	14.27	1.00
Alkaline Chlorination	11.36	1.05
Alkaline Chlorination	12.68	1.50
Electrolysis	2.38	2.67
Ozonation	4.44	3.59
Ozonation	18.49	3.43
Evaporation	2.39	9.33

Table 34. Operating Costs for Cyanide Waste Treatment.

Treatment Process	Treatment Cost ($/lb cyanide destroyed)	Reference
Alkaline Chlorination (1)	0.28–1.28	15
Alkaline Chlorination (1)	4.50	4
Alkaline Chlorination (2)	0.50–2.45	15
Kastone Process (1)	0.41	28
Electrolysis	0.082	10
Ozonation	0.14–0.64	20

Notes: (1) Oxidation to level of cyanate
(2) Complete oxidation to CO_2

REFERENCES

1. Bernardin, F. E. "Cyanide Detoxification Using Adsorption and Catalytic Oxidation on Granular Activated Carbon," *J. Water Poll. Cont. Fed.* **45** (2), 221-231 (1973).
2. Hansen, N. H. "Design and Operation Problems of a Continuous Automatic Plating Waste Treatment Plant at the Data Processing Division, IBM, Rochester, Minnesota," *Proc. 14th Purdue Ind. Waste Conf.* 227-249 (1959).
3. Palla, L. T. and R. G. Spicher. "Cyanide Treatment in Profit and Cure," Presented at 26th Ind. Waste Conf., Purdue University, 1971.
4. Watson, K. S. "Treatment of Complex Metal-Finishing Wastes," *Sewage Ind. Wastes* **26**, 182-194 (1954).
5. "Ozone Counters Waste Cyanide's Lethal Punch," *Chem. Eng.* **24**, 63-64 (1958).
6. Crowle, V. "Effluent Problems as They Affect the Zinc Die-Casting and Plating-on-Plastics Industries," *Metal Finishing J.* **17**, 51-54 (1971).
7. Lowe, W. "The Origin and Characteristics of Toxic Wastes, with Particular Reference to the Metal Industries," *Water Poll. Cont. (London)* 270-280 (1970).
8. "A State of the Art Review of Metal Finishing Waste Treatment,"

U.S. EPA Publication 12010 EIE 11/68 (1968).

9. O'Neill, F. "Facing up to Pollution," *Plating,* **57,** 1211-1213 (1970).
10. Easton, J. K. "Electrolytic Decomposition of Concentrated Cyanide Plating Wastes," *J. Water Poll. Cont. Fed.* **39,** 1621-1625 (1967).
11. American Electroplaters Society. "A Report on the Control of Cyanides in Plating Shop Effluents," *Plating* **56,** 1107-1112 (1969).
12. Beckenn, W. E. "Treatment of Cyanide Waste," *Electroplating Metal Finishing* **25** (12), 20-21 (1972).
13. Dvorin, R. "Water and Waste Treatment—A Review of Methods for the Metal Finishing Industry," *Metal Finishing Guidebook-Directory,* 28th edition (1960).
14. Crowle, V. A. "Effluent Treatment and Materials Recovery from the Metal Finishing Industry Using the Integrated Method of Treatment," *Water Poll. Cont. (London)* 636-645 (1972).
15. Zievers, J. F., R. W. Crain and F. G. Barclay. "Waste Treatment in Metal Finishing: U.S. and European Practice," *Plating* **55,** 1171-1179 (1968).
16. Besselievre, E. B. *The Treatment of Industrial Wastes,* (New York: McGraw-Hill Book Company, 1969).
17. Lavin, D. and C. Lavin. "A Practical Approach to Water Pollution Control for the Plating Company," Presented at 27th Indust. Waste Conf., Purdue University, 1972.
18. Fisco, B. "Plating and Industrial Waste Treatment at the Expanded Plant of Fisher Body, Elyria, Ohio," *Proc. 25th Purdue Ind. Waste Conf.* (1970).
19. Schink, C. A. "Plating Wastes: A Simplified Approach to Treatment," *Plating* **55,** 1302-1305 (1968).
20. Green, J. and D. H. Smith. "Processes for the Detoxification of Waste Cyanides," *Metal Finishing J.* **18,** 229-232 (1972).
21. O'Connor, S. F., B. W. Mountjoy, Jr. and N. S. Chamberlin. "Western Electric Builds Modern Plant," *Water Waste Eng./Ind.* **9,** D16-D19 (1972).
22. Beevers, M. "Chlorine and Sulfur Dioxide in the Treatment of Cyanide and Chromium Wastes," *Metal Finishing J.* **18,** 232-235 (1972).
23. Yagishita, A. "Removal of Ferri- and Ferrocyanides from Industrial Wastes," *Chem. Abstr.* **78,** 163794m (1973).
24. Curry, N. A. "Philosophy and Methodology of Metallic Waste Treatment," Presented at 27th Ind. Waste Conf., Purdue University (1972).
25. "An Investigation of Techniques for Removal of Cyanide from Electroplating Wastes," U.S. EPA Publication 12010 EIE 11/71 (1971).
26. "Electrolysis Speeds up Waste Treatment," *Environ. Sci. Technol.* **4,** 201 (1970).
27. Kandzas, P. F. and A. A. Mokina. "Use of Ozone for Purifying Industrial Waste Waters," *Chem. Abstr.* **71,** 237 (1969).
28. "New Process Detoxifies Cyanide Wastes," *Environ. Sci. Technol.* **5,** 496-497 (1971).

29. Kibbel, W. H., Jr., C. W. Raleigh and J. A. Shepherd. "Hydrogen Peroxide for Industrial Pollution Control," Presented at 27th Ind. Waste Conf., Purdue University (1972).
30. Culotta, J. M. and W. F. Swanton. "Recovery of Plating Wastes: Selection of Lowest Cost Evaporator," *Plating* **57**, 1221-1223 (1970).
31. Gallo, B. R. and J. M. Culotta. "Save on Plating Waste System," *Water Wastes Eng.* **9** (1), A18-A19 (1972).
32. Zievers, J. F. and C. N. Novotny. "Recovery of Mixed Rinse Waters by Means of Ion Exchange," *Plating* **58** (5), 482-485 (1971).
33. Golomb, A. "Application of Reverse Osmosis to Electroplating Waste Treatment, Part II: The Potential Role of Reverse Osmosis in the Treatment of Some Plating Wastes," *Plating* **59**, 316-319 (1972).
34. "Ultrathin Membranes for Treating Metal Finishing Effluents by Reverse Osmosis," U.S. EPA Publication 12010 DRH 11/71 (1971).
35. Phasey, N. W. "Experiences with Ion Exchange Resins for Effluent Treatment," *Prod. Finishing* **25** (11), 28-34 (1972).
36. Spatz, D. D. "Electroplating Waste Water Processing with Reverse Osmosis," *Prod. Finishing* **36** (11), 79-87 (1972).
37. Cheremisinoff, P. N. and Y. H. Habib. "Cyanide—An Assessment of Alternatives for Water Pollution Control," *Water Sewage Works* R95-R107 (1973).
38. Coulter, K. R. "Pollution Control and the Plating Industry," *Plating* **57**, 1197-1202 (1970).

10

TREATMENT TECHNOLOGY FOR FLUORIDE

INDUSTRIAL SOURCES

Industries that discharge significant quantities of fluorides in process waste streams include coke manufacture, glass and ceramic manufacture, transister manufacture, electroplating operations, steel and aluminum processing, and pesticide and fertilizer manufacture. Glass and plating wastes typically contain fluoride in the form of hydrogen fluoride (HF) or fluoride ion (F^-), depending upon the pH of the waste. Fluoride discharged from fertilizer manufacturing processes is typically in the form of silicon tetrafluoride (SiF_4), as a result of processing of phosphate rock (1,2). The aluminum processing industry utilizes the fluoride compound cryolite (Na_3AlF_6) as a catalyst in bauxite ore reduction. Previously, gaseous fluorides resulting from this process were discharged directly into the atmosphere. Wet scrubbing of the process fumes, an air pollution abatement procedure, now results in transfer of the fluorides to aqueous waste streams. An aluminum processing waste stream flow chart is presented in Figure 22. Average fluoride values for aluminum reduction plants are reported as 107–145 mg/l in wastewater

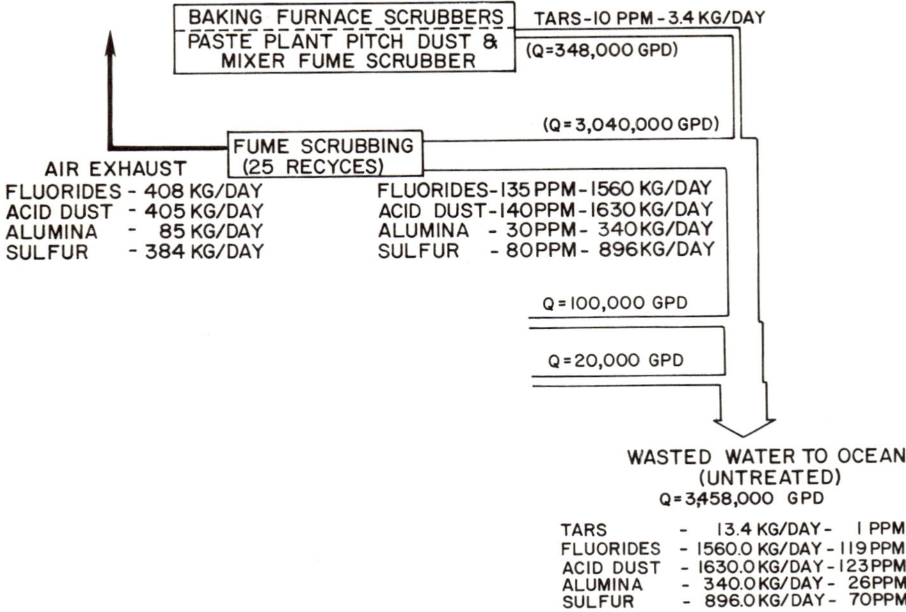

Figure 22. Aluminum reduction process waste stream flow chart, and summary of waste characteristics. [From Sylvester, *et al.* (3), courtesy of Purdue University Industrial Waste Conference.]

streams (3). Concentrations an order of magnitude greater have been reported for glass manufacturing, ranging from 1000–3000 mg/l of fluoride (1).

TREATMENT TECHNOLOGY

A variety of treatment methods are available for fluoride-bearing waste streams. A summary of the more important of these treatment processes, and their removal efficiencies is presented in Table 35. Current treatment methods can be divided into two categories; precipitation methods and adsorptive methods.

Precipitation methods involve addition of treatment chemicals, and formation of fluoride precipitates or coprecipitation of fluoride on a resulting precipitate. Removal is accomplished by solids separation of the precipitate. Treatment efficiency is, therefore, partly dependent upon the effectiveness of liquid-solid separation. Chemicals employed include lime, magnesium compounds (*e.g.,* dolomite), and aluminum sulfate (alum).

Table 35. Summary of Fluoride Treatment Processes and Levels of Treatment Achieved.

Treatment Process	Fluoride Concentration (mg/l) Initial	Final	Current Application	Ref.
Lime Addition	—	10	industrial	4
Lime Addition	1000–3000	20	industrial	1
Lime Addition	1000–3000	7–8 (after 24-hr settling)	industrial	1
Lime Addition	500–1000	20–40	industrial	5
Lime Addition	200–700	6 (16-hr settling)	industrial	6
Lime Addition	45	8	industrial	7
Lime Addition	4–20	5.8 (ave.)	industrial	8
Lime Addition	590	80	industrial	8
Lime + Calcium Chloride	—	12	industrial	9
Lime + Alum	—	1.5	industrial	10
Alum	3.6	0.6–1.5	municipal	11
Alum	60	2	lab scale	9
Hydroxylapatite Beds				
synthetic	12–13	0.5–0.7	municipal	12
synthetic	10	1.6	municipal	13
bone char	6.5	1.5	municipal	14
bone char	9–12	0.6	municipal	15
Alumina Contact Beds	8	1	municipal	14
Alumina Contact Beds	9	1.3	industrial (lab scale)	1
Alumina Contact Beds	20–40	2–3	industrial (pilot scale)	5

Lime

Direct addition of lime is the standard technique for reduction of high concentrations of fluoride ion. The lime reacts with fluoride in the wastewater to produce calcium fluoride. Calcium fluoride has a theoretical maximum solubility of approximately 8 mg/l of fluoride at stoichiometric concentrations of calcium (34 mg/l). Thus, concentrations of calcium fluoride above this solubility limit will form a precipitate. Reported fluoride removals down to residuals of only 10–20 mg/l reflect the slow rates of precipitate formation (1,4). However, 24-hour contact with lime has been reported to reduce fluoride to near the theoretical limit of 8 mg/l (1). At very high lime concentrations, improved fluoride removal has been reported (Table 35). Successful lime treatment may require addition of up to 4 g hydrated lime per liter of wastewater, resulting in an effluent pH above 12 (6,16). Lime dosing at this level has been reported to reduce fluoride from 200 to 3 mg/l, in 19-hours reaction time (6). The use of a mixture of calcium chloride and hydrated lime has been used successfully at one plant, to reduce total fluoride levels to approximately 12 mg/l, after neutralization, clarification and filtration, at pH 7 to 8 (9).

In one case, coprecipitation effects may have contributed to reduction of fluoride (6). A television tube manufacturing plant combined lead precipitation with fluoride precipitation in a sequential treatment step, followed by combined sedimentation. Lead wastes were treated with trisodium phosphate, followed by introduction of the fluoride waste stream, lime and polyelectrolyte addition. Sixteen-hours sedimentation produced a 6 mg/l fluoride effluent. Addition of excess lime further reduces fluoride concentration, at the expense of additional treatment chemical costs and increased pH and calcium concentration in the waste effluent. Recarbonation of the effluent provides a means of excess calcium removal, and pH adjustment. An alternate explanation of the improved treatment observed in this instance may be the presence of phosphate, added in the lead treatment step. In two separate studies, the addition of sodium hexametaphosphate with alum has been reported to enhance treatment above that observed for alum alone (9,17). In one application of alum plus sodium hexametaphosphate (4:1 dosage ratio), an initial fluoride level of 20 mg/l was reduced to below 2 mg/l at pH 7 (9).

The lime treatment process for fluoride waste is composed of standard unit operations of the sort employed in lime softening of municipal water. These units include a rapid mixing basin for chemical addition, a flocculating unit, and a settling basin for solids separation.

One very concentrated fluoride waste (40,000 mg/l fluoride in 30% sulfuric acid) was treated by pumping the 240-gal batch waste into a 20% lime slurry, and discharging the resultant 15% solids sludge to a landfill (18). This operation was conducted three to five times per week and employed 3.33 lb of lime per gallon of waste.

Treatment of individual fluoride wastewaters prior to mixing of waste streams can greatly enhance fluoride discharge levels. Two South African steel mills of total waste flows 1 and 2.5 MGD have reported effluent levels of 2–2.5 and 1.0 mg/l fluoride respectively (19). Fluoride, originating from electrolytic tinning units, is precipitated with lime and the resulting effluent is then diluted upon combination with the total plant effluent.

Treatment costs specifically for fluoride removal are frequently difficult to isolate, as fluoride is not the only wastewater contaminant being removed and individual unit costs are not reported.

General sludge handling techniques should apply to the precipitated calcium fluoride sludge. Dewatering may be accomplished by one of the standard processes of sand bed drying, lagooning, vacuum filtration or centrifugation, depending upon availability of land and the overall nature of the final sludge (*i.e.*, other sludges combined). General costs for sludge dewatering (20) are:

Lagooning	\$1–5/dry ton
Sand Bed Drying	\$3–20/dry ton
Vacuum Filtration	\$8–50/dry ton
Centrifugation	\$5–35/dry ton

Land disposal is estimated to cost an additional \$1–5/dry ton.

Magnesium

Early work in the treatment of drinking waters containing excessive fluoride concentrations demonstrated that fluoride ion was removed during the standard lime-soda ash softening process. Removal was in direct proportion to the amount of magnesium hardness removed, and was attributed to adsorption of fluoride ion onto the magnesium hydroxide floc formed in the softening process. Fluoride concentrations of 3–4 mg/l were reported reduced to 0.8–1.12 mg/l (11). However, in the absence of appreciable magnesium occurring naturally in the water or wastewater to be treated, the requirement for supplmentary addition of magnesium may make this process economically unfeasible. No fluoride removal plant has been specifically designed to use this process. Fluoride reduction is, however, accomplished in conjunction with magnesium hardness removal at several municipal water treatment plants in the United States where fluoride is present in the raw water at concentrations of 2–3 mg/l (21).

Aluminum Sulfate (Alum)

For a water of low hardness containing 3.6 mg/l fluoride ion, simple alum addition plus small quantities of lime for pH control resulted in fluoride reduction to 1 mg/l (11). For hard waters, other methods of removal were more effective.

Fluoride removal by alum addition is essentially a coprecipitation phenomena, whereby fluoride is removed with aluminum hydroxide precipitate. Best removal appears to occur at pH 6–7, with initial fluoride about 100 mg/l reported reduced first to 5–5.2 mg/l by lime precipitation (pH 11.9) and to 1.7 mg/l by alum coagulation at pH 5.9–7 (16).

Recent studies have indicated that final fluoride level is independent of initial fluoride concentration up to an initial fluoride level of at least 100 mg/l, and is dependent only upon alum dosage and treatment pH (9). Best treatment is reported to occur at pH 7, with fluoride reduced to 4.8 mg/l at an alum dosage of 40 per pound of fluoride. At alum dosage above 200 lb per lb of fluoride, effluent fluoride levels below 2 mg/l were achieved. Even at more moderate alum dosages however, sludge volumes near 40% of the original wastewater volume treated result (9).

Fluoride removal in conjunction with arsenic treatment of a rare-earth industrial waste was reported to produce an effluent fluoride concentration of 1.5 mg/l (10). Initial lime precipitation was followed by further addition of lime, plus superphosphate and aluminum sulfate.

Adsorption

Adsorption methods involve the passage of the wastewater through a contact bed, with fluoride being removed by general or specific ion exchange or chemical reaction with the solid bed matrix. Although solids removal is not required for these operations, bed regeneration and subsequent treatment of the concentrated regenerant are an integral part of the overall treatment process.

Since the basic mechanism of a contact process is one of ion exchange or surface reaction, these methods are usually appropriate only for low level fluoride wastes, or polishing processes after fluoride reduction by previous treatment to the 10–20 mg/l level. Otherwise, the requirement for frequent bed regeneration makes the process economically unfeasible. Bed media that have been used are hydroxylapatite (synthetic, processed bone and bone char), ion exchange resins, and activated alumina.

Hydroxylapatite. Reduction of fluoride from 12–13 mg/l to 0.5–0.7 mg/l was reported, upon passage through a full-scale contact bed of synthetic hydroxylapatite (12). Later reports (21) indicate that this unit and others employing the same process were ultimately abandoned due to high chemical costs and bed attrition (up to 42% loss/year). For a similar pilot plant-scale operation in South Africa, 10 mg/l fluoride could be reduced to 1.5 mg/l (13). The presence of chlorides in the water reduced filter bed capacity and increased regeneration chemical requirements.

Natural hydroxylapatite in the form of bone char was reported capable of reducing 10 mg/l fluoride to 0.6 mg/l, but regeneration with phosphoric acid and sodium hydroxide resulted in excessive chemical costs (15).

Ion Exchange. Natural and synthetic zeolite ion exchangers have been examined and reported to be effective for removal of fluoride. The exchangers were both pretreated and regenerated with aluminum salts. Further investigation indicated that fluoride removal was attributable to aluminum oxide which precipitated in the column bed (14,22). A strong anion exchange resin selective for fluoride was developed by Rohm and Haas (14) but overall treatment and chemical costs have been reported as excessive ($0.85/1000 gal, for chemical regenerant only).

Alumina. Contact beds of activated alumina have been used for many years in municipal water treatment plants for removal of fluoride ion. One unit, in Bartlett, Texas, has operated successfully since 1952 using a 500-cu ft bed regenerated by sodium hydroxide and neutralized with sulfuric acid. Fluoride is reduced from 8 mg/l to 1 mg/l (14). The Bartlett, Texas, plant costs were $0.052/1000 gal, based on 1951 cost data. Equipment costs were $11,360 and bed media $4000. The building housing the plant was provided at no cost. Water was treated at the rate of 400 gpm. Bellock (23) reported operating costs for fluoride and arsenic removal with activated alumina columns to range from $0.015–0.05/1000 gal.

Use of an alumina bed as a polishing unit to follow lime precipitation of high fluoride (1000–3000 mg/l) wastes resulted in 30 mg/l residual fluoride carryover from the precipitation process being reduced to approximately 2 mg/l upon passage through the contact bed (1). At a pH of 11.0 to 11.5, a concentration of 9 mg/l fluoride was reduced to 1.3 mg/l. In regenerating, 4% of the alumina was lost for every 100 regeneration cycles.

A proprietary package electrochemical process has been developed which claims to reduce fluoride to 1.5 mg/l or less (24). The system operates as continuous flow and utilizes consumable electrodes. No pH adjustment or chemical additions are required, and economics are reported comparable to precipitation and adsorption treatment methods.

SUMMARY

Industrial wastes containing high fluoride levels often require two-stage treatment. Lime precipitation removes fluorides down to 10 mg/l or less. Further reductions can be accomplished down to the 1 mg/l level with other existing techniques. The preferred method seems to be activated alumina, with alum regeneration. The units involved are essentially those used for ion exchange water softening, and the regenerant can be treated in the high lime removal unit by recycling. Although treatment levels obtainable by the various fluoride removal techniques are well defined in the literature, there is much less information available on costs. Published data, however, indicate that operating costs are generally in the range of $0.10–0.50/1000 gal.

REFERENCES

1. Zabban, W. and H. W. Jewett. "The Treatment of Fluoride Wastes," *Proc. 22nd Purdue Industrial Waste Conf.* **22,** 706-716 (1967).
2. Cherry, J. M. "Fluorine Recovery in Phosphate Manufacture," *Water Wastes Eng.* **7,** D-5 (1970).
3. Sylvester, R. O., R. T. Oglesby, D. A. Carlson and R. F. Christman. "Factors Involved in the Location and Operation of an Aluminum Reduction Plant," *Proc. 22nd Purdue Industrial Waste Conf.* 441 (1967).
4. Schink, C. A. "Plating Wastes: A Simplified Approach to Treatment," *Plating* **55,** 1302-1305 (1968).
5. Petrova, T. N. and U. G. Bakhurov. "Purification of Waste Solutions from Fluorine," *Atomn. Eng.* **26,** 552-553 (1969); *Water Poll. Abstr.* **43,** No. 609 (1970).
6. Rohrer, K. L. "An Integrated Facility for the Treatment of Lead and Fluoride Wastes," *Ind. Wastes* **17** (9), 36-39 (1971).
7. "Pulling Effluents Into Line," *Chem. Eng.* **78,** 40 (1971).
8. Teer, E. H. and L. V. Russel. "Heavy Metals Removal from Wood Preserving Wastewater," Presented at the 27th Ind. Waste Conf., Purdue University, 1972.
9. Zabban, W. and R. Helwick. "Defluoridation of Wastewater," Presented at 30th Annual Purdue Ind. Waste Conf., 1975.
10. Skripach, T., V. Kagan, M. Ramanov, L. Kamin and A. Semina. "Removal of Fluorine and Arsenic from the Waste Water of the Rare-Earth Industry," *Proc. 5th Internat. Conf. Water Poll. Res.* (1970), 2; Pap-No III—34 (1971); *Water Poll. Abstr.* **564,** No. 2770 (1971).
11. Culp, R. L. and H. A. Stoltenburg. "Fluoride Reduction at LaCrosse, Kansas," *J. Amer. Water Works Assoc.* **50,** 423-431 (1958).
12. Wamsley, R. and W. F. Jones. "Fluoride Removal," *Water Sewage Works* **94,** 372-376 (1947).
13. Cillie, G. G., O. O. Hart and G. J. Standy. "Defluoridation of Water Supplies," *J. Inst. Water Eng.* **12,** 203-210 (1958).
14. Maier, F. J. "Defluoridation of Municipal Water Supplies," *J. Amer. Water Works Assoc.* **45,** 879-888 (1953).

15. "Treating a High Fluoride Water," *Public Works* **86,** 67 (1955).
16. Rabosky, J. G. and J. P. Miller. "Fluoride Removal by Lime Precipitation and Alum and Polyelectrolyte Coagulation," *Proc. 29th Ind. Waste Conf.,* Purdue University (1974).
17. Miller, D. G. "Fluoride Precipitation in Metal Finishing Waste Effluent," Water-1974, I. Industrial Wastewater Treatment, *AIChE Symposium Series* No. 144, Vol. 70 (1974).
18. Lewder, L. R. "Fluoride Waste Puzzle Solved," *Water Wastes Eng.* **8,** B-6 (1971).
19. Heynike, J. J. C. and F. U. K. von Reiche. "Water and Pollution Control in the Iron and Steel Industry with Special Reference to the South African Iron and Steel Industrial Corporation," *Water Poll. Cont. (London)* 569-673 (1969).
20. Burd, R. S. "A Study of Sludge Handling and Disposal," Publication No. WP-20-4, (Washington, D.C.: U.S. Dept. of the Interior, 1968).
21. Savinelli, E. A. and A. P. Black. "Defluoridation of Water and Activated Alumina," *J. Amer. Water Works Assoc.* **50,** 33-44 (1958).
22. Maier, F. J. "Water Defluoridation at Britton: End of an Era," *Public Works* **102,** 70 (1971).
23. Bellock, E. "Arsenic Removal from Potable Water," *J. Amer. Water Works Assoc.* **63,** 454 (1971).
24. Duffey, J. G. "Andco Fluoride Removal System," *Tech. Bull.,* Andco Environ. Processes, Inc. Buffalo, N.Y.

11

TREATMENT TECHNOLOGY FOR IRON

INDUSTRIAL SOURCES

Industrial sources of iron include (1-3):

Mining operations	Textile mills
Ore milling	Food canneries
Chemical industries (organic, inorganic, petrochemical)	Tanneries
	Titanium dioxide production
Dye industries	Petroleum refining
Metal processing industries	Fertilizers

Iron exists in the ferric or ferrous form, depending upon conditions of pH and dissolved oxygen concentration. At neutral pH and in the presence of oxygen, soluble ferrous iron (Fe^{+2}) is oxidized to ferric iron (Fe^{+3}), which readily hydrolyzes to form the insoluble precipitate, ferric hydroxide, $Fe(OH)_3$. At pH below 6, however, the rate of ferrous iron oxidation to the ferric form is extremely slow. Therefore, acidic and/or anaerobic conditions are necessary for appreciable concentrations of soluble iron to exist. In addition, at pH values above 12, ferric hydroxide will resolubilize due to the formation of the $Fe(OH)_4^-$ anion. Ferrous and ferric iron may also be solubilized in the presence of cyanide, due to the formation of ferro- and ferri-cyanide complexes. Such species present considerable difficulties for both iron and cyanide treatment.

Perhaps the most significant industrial source of soluble iron waste is spent acid pickling solution. Pickling baths are employed to remove oxides from iron and steel during processing, or prior to plating of other metals on the surface. These baths contain strong acid solutions; usually sulfuric acid of 5–20% concentration by weight, although more recently hydrochloric acid has come into wide use. The baths accumulate soluble iron until its concentration interferes with product quality. At that time the bath must be replaced.

Acid mine drainage waters are also high in soluble iron and represent a formidable treatment problem. It has been reported that 75% of the mines producing these wastes are inactive and abandoned (4). In one area, the flow from abandoned mines represents 90% of the total acid mine drainage of the region (5). Typical iron concentrations reported for industrial wastewaters and mine waters are given in Table 36.

TREATMENT TECHNOLOGY

Oxidation—Precipitation

The predominant treatment process for iron is conversion of ferrous iron to the ferric state, and precipitation of ferric hydroxide near pH 7, where its solubility is at a minimum. Conversion of ferrous to ferric iron occurs readily upon aeration at pH 7.0–7.5. Spontaneous formation of ferric hydroxide then results in iron removal by precipitation. The rate of oxidation of ferrous to ferric iron is dependent upon pH, buffer intensity in bicarbonate systems, and upon the concentration of dissolved organic matter (22). Those organic compounds, which contain hydroxyl and carbonyl functional groups, seem to greatly reduce oxidation rates and can result in high soluble iron levels if the organic substances do not readily break down. If iron is present in acid waters as the ferric ion, simple neutralization suffices to effect iron removal.

Iron removal from municipal water supplies has been accomplished successfully by the water treatment industry for many years. Iron levels in ground water supplies are characteristically in the 1–10 mg/l range, as ferrous iron. Ground waters are typically anoxic and weakly acid, usually high in CO_2. Aeration to remove CO_2 (and consequently increase pH) and to oxidize ferrous to

Table 36. Iron Concentrations Reported for Industrial Wastewaters.

Source	Soluble Iron Concentration (mg/l)	Reference
Mine waters		
Mine Drainage	360 (36)[a]	6
Mine Drainage	10–200, 93 ave	4
Mine Drainage	17–218	4
Mine Drainage	291–800	7
Trailings pond	3,200	4
Acid Mine water	501	8
Mine waste	67–70 (64–67)[a]	9
4 coal mines	122–330 (8–313)[a]	10
3 coal mines	40–150 (21–150)[a]	11
Motor Vehicle Assembly		
Body assembly	4	12
Vehicle assembly	3	12
Steel Processing		
Waste pickle liquor	96,800	13
Waste pickle liquor	70,000	14
Pickle bath rinse	200–5,000	15
Pickle bath rinse	60–1,300, 210 ave	16
Steel cold finishing mills	60–150	17
Metal Processing		
Appliance manufacturer	0.09–1.9	18
Automobile heating controls	1.5–31	19
Appliances		
Mixed wastes	0.2–20	20
Spent acids	25–60	
Chrome plating	40	14
Plating wastes	2–4	21

[a] Values in parentheses are for ferrous iron levels.

Table 37. Summary of Treatment Performance for Municipal Iron Treatment Plants.

Process	Iron Concentration, mg/l		Reference
	Influent	Effluent	
1. Chlorination, alum-lime-sodium silicate precipitation, sand filtration	1.5	0.05	23
2. Aeration, lime, sand filtration	2.4	0.0	23
3. Aeration, sand filtration	2.5	0.13	25
4. Aeration, coke bed filtration, sedimentation, sand filtration	5.7	0.13	26
5. Lime, aeration, diatomite filtration	10	0.1	27

ferric iron comprises the standard municipal water treatment. Lime is frequently employed for additional pH adjustment. Table 37 summarizes treatment efficiencies obtained from full-scale municipal water treatment plants.

Iron treatment processes for most industrial wastes differ from that of the water industry, because for iron removal it is often necessary to neutralize fairly high concentrations of acid contained in the iron-bearing wastewaters. Neutralization increases treatment costs considerably, and since lime is the preferred base and sulfuric acid is generally present in the waste, large quantities of insoluble gypsum (calcium sulfate) are produced, which complicates the sludge disposal problem. The use of limestone slurries may offer advantages with respect to solids removal and handling. These advantages may not necessarily always offset the cost differences and treatment levels (Table 38).

Iron bearing wastes in industry frequently are the result of acid treatment of iron and steel to remove surface oxides. Two types of waste are generated by this cleaning process, referred to as "pickling." Rinse waters may typically contain 50–1500 mg/l of ferrous iron and considerable acid (see Table 36). The pickling baths themselves are extremely concentrated in acid and, ultimately, soluble iron. This necessitates periodic dumping of the bath, which can represent a flow of several tens of thousands of gallons.

Treatment of waste pickle liquor is normally best handled separately from rinse waters (17). Cost of rinse water treatment will be a function of many variables including: plant location, extent of waste stream segregation, total flow, particular waste characteristics, auxiliary facilities required and the effluent standards in force. Over the 10-year period 1960-1970, costs for such plants ranged from \$208–\$625/1000 gpd (median \$450/1000 gpd) for plants with flows of 14.4–21.6 MGD (17). These costs excluded collection system costs, which can be significant. Corresponding operating costs were

Table 38. Treatment of Acid Mine Drainage (5).

Treatment Chemical	Iron (mg/l) Initial	Iron (mg/l) Final	Percent Removal	Chemical Cost (¢/1000 gal)	Cost per mg Acidity per MG (¢)
Limestone slurry	145	5	96.2	5.9	10
Lime	157	1.6	99.0	3.2	5.2
Soda Ash	126	2.3	98.2	26.8	4.9

7–25¢/1000 gal. Plants generally incorporated the following steps:

1. Oil treatment
2. Chemical neutralization—flocculation/clarification
5. Chrome treatment
4. Deep well injection for waste pickle and chrome baths
5. Sludge centrifugation and landfill disposal

In general, one of the most costly processes involving iron removal is treatment of the waste pickle liquor itself. In addition to high iron levels and ferrous to ferric ratios, the waste is strongly acidic. Treatment cost is more directly related to neutralization of acidity than to iron removal. Capital and operating costs for a typical acid neutralizing plant are given in Table 39, for various waste flows and acid concentrations.

Table 39. Acid Neutralization Costs, Including Equalization and Sludge Dewatering (2).

Flow (MGD)	Acidity (mg/l)	Capital Cost ($1000)	Operating Cost[a] (¢/1000 gal)
0.5	500	145.0	15
0.5	1,000	190.0	29
0.5	20,000	514.0	108
1.0	500	236.0	25
1.0	1,000	320.0	26
1.0	20,000	807.0	103
10.0	500	1,290.0	12.5
10.0	1,000	1,640.0	20
50.0	500	3,980.0	11

[a] Costs based upon a suspended solids concentration of 500 mg/l.

Capital costs are also presented in Figure 23, exclusive of treatment facilities and sand filtration (2). As acid levels in the waste increase, capital costs go up. Figure 24 illustrates this relationship for a 1-MGD plant.

Combined operating costs for neutralization plus precipitated sludge dewatering, as a function of acidity, are illustrated in Figure 25. To achieve low-effluent iron levels, it is frequently necessary to pass the clarifier effluent through a sand or mixed media filtration unit. The capital and operating costs of such filters are presented in Figure 26. An operating cost of 2.3¢/1000 gal has been reported for mixed media filtration in municipal tertiary treatment, for a 7.5–MGD facility (28). Costs of $21–$35/1000 gpd and 1/2¢/1000 gal have been cited for high-rate mixed media filters (16 gpm/ft^2)

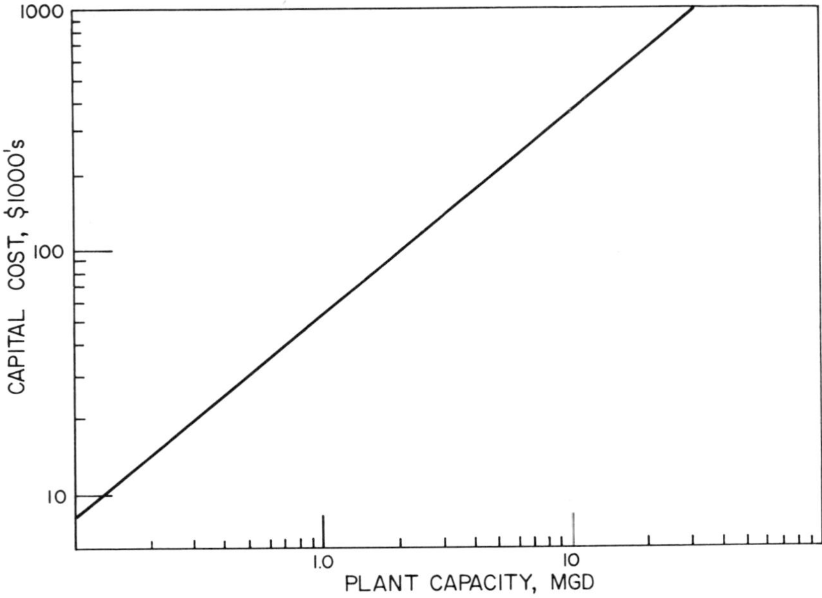

Figure 23. Capital costs for an acid waste neutralization facility, excluding sludge treatment facilities (2).

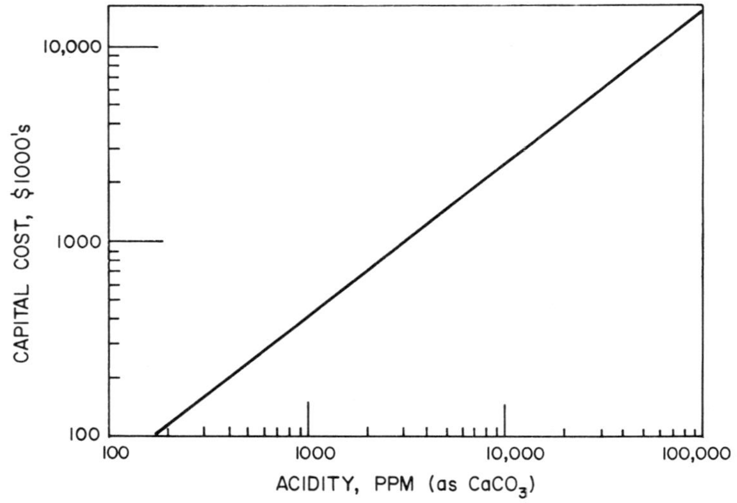

Figure 24. Capital costs for lime neutralization of acid wastes. Cost vs. wastewater acidity (2).

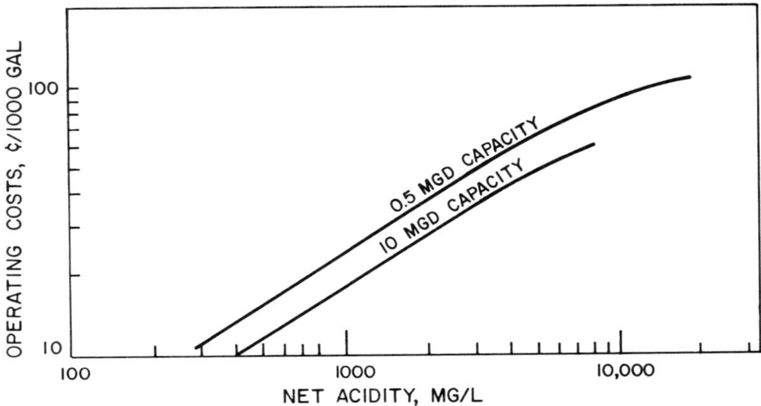

Figure 25. Operating costs for lime neutralization of acidic wastes, including sludge dewatering by vacuum filtration (2).

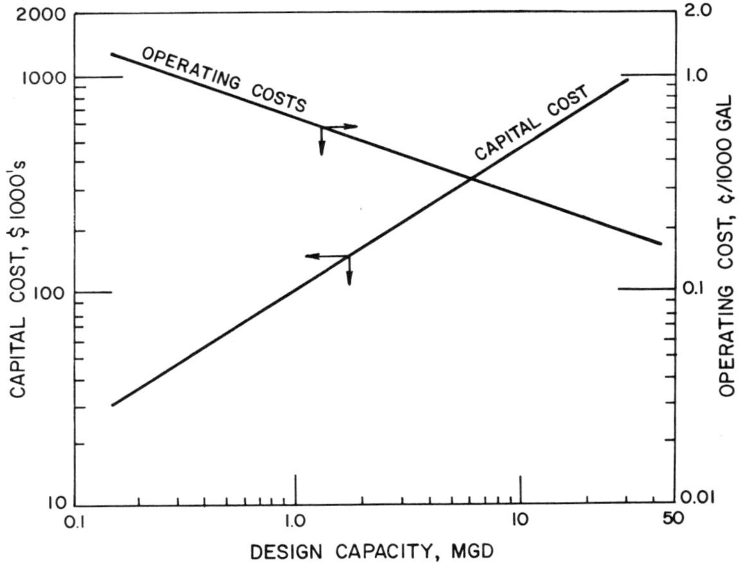

Figure 26. Capital and operating costs of filtration through sand or graded media (2).

at flows of 14.4 MGD (29). Solids disposal has not been considered in these data.

Waste pickle liquor is the most difficult and expensive iron-bearing waste to treat on a per unit volume basis. Neutralization and subsequent precipitation with lime has been most commonly used by steel mills to treat these wastes. Treatment costs are summarized in Table 40.

Rinse waters from pickle lines are also treated by neutralization and aeration. The extent of conversion of ferrous iron to ferric iron may influence the final soluble iron concentration. Generally, the limiting factor as regards effluent iron concentration is solids removal efficiency. Limestone neutralization and aeration of an acid mine water decreased 360 mg/l soluble iron (36 mg/l Fe^{+2}, 324 mg/l Fe^{+3}) to 7 mg/l upon primary settling (6). Additional settling overnight reduced the residual iron to essentially zero.

The use of limestone in treating of pickling rinse waters has been practiced at full-scale operation (16,33). Rinse waters ranging from 600–13,000 mg/l iron (ave 210) with corresponding acidity of 80–5800 mg/l as $CaCO_3$ (ave 490) was reduced to 4.0 mg/l iron on the average (range 0.32–23 mg/l) with limestone slurry in 50% excess, followed by aeration. Capital costs for the 2.16-MGD facility was $630/1000 gpd and operating costs, exclusive of sludge handling or supervision, were 24¢/1000 gal or $4.38/ton of steel pickled. The unit employs sludge recirculation and vacuum filtration. Since the system uses hydrochloric acid pickle, the effluent would be expected to contain high soluble calcium and chloride concentrations.

Acid mine drainage is similar to steel industry rinse wastes. Cost data for various industrial and mine water treatments are given in Table 41.

Frequently, two waste streams can be combined to provide treat-

Table 40. Cost of Waste Pickle Liquor Treatment.

Treatment	Plant Capacity (MGD)	Capital Cost ($/1000 gpd)	Operating Cost ($/1000 gal)	Reference
Neutralization, precipitation and sludge landfill	na	na	$30–40	13
Neutralization and precipitation	na	na	$20	30
Neutralization, precipitation and sludge lagooning	na	na	$10–40	31
Neutralization, precipitation and sludge disposal	0.06	$1667	$20	32

ment for one or both. The most obvious example is neutralization by combination of acidic and basic solutions. Occasionally, chromate wastes may be present in the same plant with waste pickle liquor or rinse water containing appreciable ferrous iron. In place of separate treatment, the pickle liquor can be employed as a reductant for chromate, followed by coprecipitation of chromium and iron hydroxide by lime addition. One such process has been reported to reduce iron and chromium in the final effluent to 0.5 and 0.05 mg/l respectively (34).

Deep Well Disposal

Deep well injection has become increasingly popular with steel companies as a disposal technique, particularly for waste pickle liquor (31). Costs are widely variable depending upon depth of well and injection pressure requirements. Wells may cost $50,000–75,000 to drill and case, and operating costs may be of the order of $2.50–$10.00/1000 gal (13). Another report (35) indicates that capital costs may vary from $30,000 for a 1800-ft deep well where no surface equipment is required, to $1,400,000 for another of 12,000-ft depth, plus a clarifier, dual filters and four displacement-type pumps. The filter requirement applies to any waste containing suspended solids, as they clog the well. Figure 27 and 28 present

Table 41. Cost Summary for Treatment of Industrial and Acid Mine Iron-Bearing Waters by Neutralization and Precipitation.

Waste	Flow (MGD)	Capital Costs ($/1000 gpd)	Operating Costs (¢/1000 gal)	Reference
Pickling Rinse Water	2.16	630	24	16
Steel Cold Finishing				
Mill Wastes[a]	14.4–21.6	208–625	7–25	17
Acid Mine Drainage[a]				11
(Limestone)				
90 mg/l Fe^{+2},				
190 mg/l acidity	4.0	165	9.7	
500 mg/l Fe^{+2},				
1000 mg/l acidity	0.1	1303	66	
	1.0	352	27	
	7.0	194	22	
5000 mg/l Fe^{+2},				
8000 mg/l acidity	0.1	1732	213	
	1.0	765	171	
	7.0	556	166	

[a] Includes sludge disposal costs.

capital and operating costs for deep well injection systems. Deep well injection is not feasible for rinse water disposal, due to the large waste volume involved.

Other Methods

Neutralization of waste pickle liquor is uneconomical and often impractical, due to the quantities of sludge produced. Recovery processes for sulfuric acid pickle liquor and for hydrochloric acid pickle liquor are available (36). Various pickle liquor recovery processes are in operation, and their nature and some cost data have been summarized (37). For each process available, however, the primary emphasis is on recovery of acid rather than removal of iron. In brief, the processes evaporate the waste liquor and remove iron as ferrous oxide, ferrous sulfate or ferric oxide, and acid as

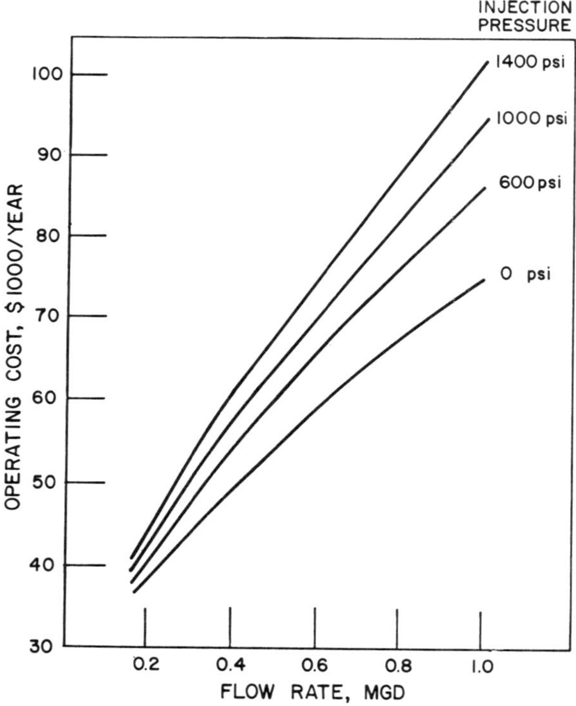

Figure 27. Annual operating costs of waste disposal by deep well injection (2).

either concentrated hydrochloric or sulfuric acid.

Jackson (38) has described a recovery process for hydrochloric acid pickle liquor, which can convert the ferrous chloride to free iron or ferric oxide, while regenerating the hydrochloric acid. A flow chart of a commercial unit in operation in Canada is presented in Figure 29. This unit employs reduction of the iron by hydrogen gas. Although the original unit was designed for use with scrap, the application to pickle liquor is direct, as indicated in Figure 29. The system is claimed to be capable of successful and economical operation.

Reverse osmosis has been suggested for the treatment of acid mine drainage. A pilot plant reverse osmosis unit was operated on-site for four months, and reported iron reduction from 67 ± 5 mg/l to 0.60 ± 0.06 mg/l total iron and 64 ± 4 mg/l to 0.57 ± 0.06 mg/l ferrous iron (9). The concentrated brine was then subjected to neutralization and precipitation, with reported effluent concentrations from this process of 0.9–6.0 mg/l. Projected costs for a 0.50-MGD system were given as $607/1000 gpd capital and 48.5¢/1000 gal operating.

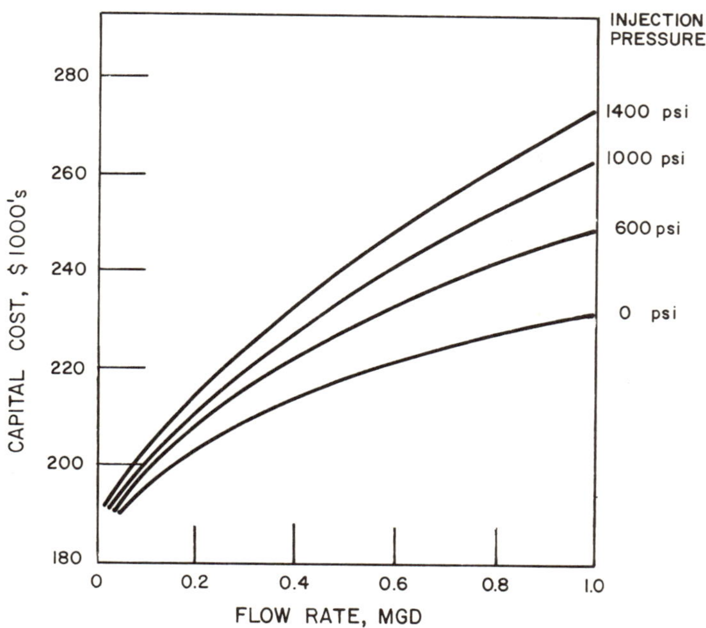

Figure 28. Capital costs for deep well injection systems (2).

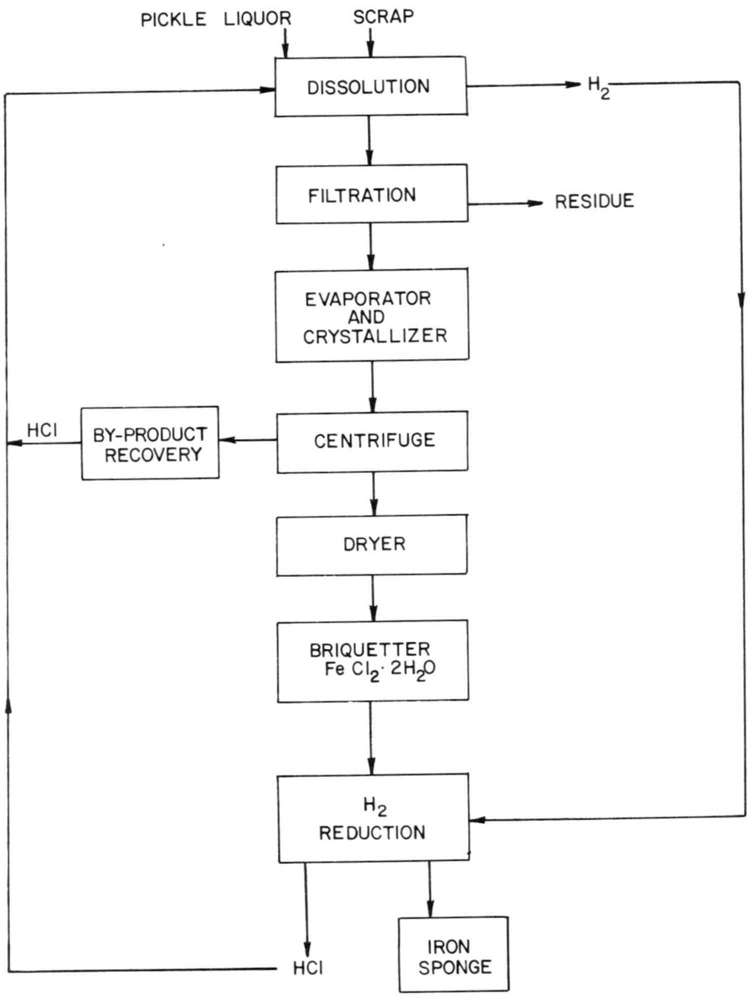

Figure 29. Outline flowsheet of the Peace River Mining and Smelting Co. process for treatment of ferrous scrap (39).

SUMMARY

Existing technology is capable of soluble iron removal to levels below 0.3 mg/l. Failure to achieve these levels appears to be the result of improper pH control resulting in increased solubility of ferric hydroxide during coagulation, incomplete oxidation of the more soluble ferrous iron to the ferric state, or poor solids removal. Apparent soluble iron may actually be present in effluent waters as the finely divided colloidal state, due to inefficient removal of the ferric hydroxide. Precipitation treatment systems which consistently produce low-iron waters also employed some polishing treatment, such as rapid sand filters.

Oxidation and precipitation is capable of providing effective iron removal from most wastes, providing that prior treatment has removed iron complexing agents (*e.g.*, cyanides) prior to precipitation, and that the waste pH is suitable. However, treatment costs may be extremely high for chemical requirements and sludge dewatering and disposal, particularly when wastes contain high concentrations of sulfuric acid, due to precipitation of calcium sulfate sludge. Although dollar values are high, estimates for waste treatment facilities range from 5–15% of the installation and operating costs of the process.

Recovery processes appear to be currently economical only for large industries such as steel mills, although certain process modifications may make recovery more feasible and enhance treatment efficiencies at reduced costs. Other than precipitation and recovery, only deep well disposal seems economically competitive for industries with high acid, high metal waste solutions. The costs, if geological conditions are favorable, are likely to be more attractive to smaller industries than other treatment alternatives.

REFERENCES

1. McKee, J. E. and H. W. Wolf. *Water Quality Criteria,* 2nd ed., California State Water Quality Board Publication, No. 3-A (1963).
2. *The Economics of Clean Water, Vol. III, Inorganic Chemical Industry Profile* (Washington, D.C.: U.S. Dept. of the Interior, 1970).
3. Dean, J. C., F. L. Bosqui and K. H. Lanouette. "Removing Heavy Metals from Waste Water," Environ. Sci. Technol. **6**, 518-522 (1972).
4. Hill, R. D. "Control and Prevention of Mine Drainage," presented at the Environmental Resources Conference on Cycling and Control of Metals, Battelle Columbus Laboratories, Columbus, Ohio, 1972.
5. Wilmoth, R. C. and R. D. Hill. "Neutralization of High Ferric Iron Acid Mine Drainage," U.S. EPA Report 14010 ETV 8/70 (1970).
6. Mihok, E. A., M. Duel, C. E. Chamberlain and J. G. Selmeczi. "Mine Water Research. Limestone Neutralization Process," U.S. Bur. Mines, Rep. Invest. No. 7191 *Chem. Abstr.* **70** (6443), p. 1969.
7. Larsen, H. P., J. K. P. Shou and L. W. Ross. "Chemical Treatment

of Metal-Bearing Mine Drainage," *J. Water Poll. Cont. Fed.* **45** (8), 1682-1695 (1973).

8. Kunin, R. and D. G. Downing. "Ion Exchange System Boasts More Pulling Power," *Chem. Eng.* **79**, 67-69 (June 28, 1971).
9. Rex Chainbelt, Inc. "Reverse Osmosis Demineralization of Acid Mine Drainage," U.S. EPA Report 14010 FQR, 03/72 (1972).
10. Bituminous Coal Research, Inc. "Optimization and Development of Improved Chemical Techniques for the Treatment of Coal Mine Drainage," U.S. EPA Report 14010 EIZ, 01/70 (1970).
11. Bituminous Coal Research, Inc. "Studies of Limestone Treatment of Acid Mine Drainage. Part II," U.S. EPA Report EIZ 12/71 (1971).
12. *The Cost of Clean Water, Vol. III, Industrial Waste Profiles, No.* **2**, *Motor Vehicles and Parts,* (Washington, D.C.: U.S. Dept. of the Interior, 1967).
13. Southern Research Institute. "An Electromembrane Process for Regenerating Acid from Spent Pickle Liquor," U.S. EPA Report 12010 EQF 3/72 (1972).
14. Parsons, W. A. *Chemical Treatment of Sewage and Industrial Wastes,* (Washington, D.C.: National Lime Association, 1965).
15. Nemerow, N. L. *Theories and Practice of Industrial Waste Treatment,* (Reading, Massachusetts: Addison-Wesley Publishing Co., 1963).
16. Armco Steel Corporation. "Limestone Treatment of Rinse Waters from Hydrochloric Acid Pickling of Steel," U.S. EPA Report 12010 DUL 02/71 (1971).
17. Donovan, E. J., Jr. "Treatment of Wastewater for Steel Cold Finishing Mills," *Water Wastes Eng.* **7**, F22-F25 (November, 1970).
18. Watson, K. S. "Treatment of Metal Finishing Wastes," *Sewage Ind. Wastes,* **26**, 182-194 (1954).
19. Gard, C. M., C. A. Snavley and D. J. Lemon. "Design and Operation of a Metal Works Waste Treatment Plant," *Sewage Ind. Wastes* **23**, 1429-1439 (1951).
20. Anderson, J. S. and E. H. Iobst, Jr. "Case History of Wastewater Treatment in a General Electric Appliance Plant," *J. Water Poll. Cont. Fed.* **40**, 1786-1795 (1968).
21. Battelle Columbus Laboratories. "An Investigation of Techniques for Removal of Cyanide from Electroplating Wastes," U.S. EPA Report 12010 EIE 11/71 (1971).
22. Jobin, R. and M. M. Ghosh. "Effect of Buffer Intensity and Organic Matter on the Oxygenation of Ferrous Iron," *J. Amer. Water Works Assoc.* **64**, 590-595 (1972).
23. Haney, C. G. "Potassium Permanganate Solves a Problem at Charlottesville, Va.," *Water Sewage Works* **111**, R353-R354 (1964).
24. Owens, L. V. "Iron and Manganese Removal by Split Flow Treatment," *Water Sewage Works* **110**, R76-R86 (1963).

25. Gardner, J. R. and A. J. Stika. "Valveless Filter Installed for Iron Removal," *Public Works* **95**, 109-110 (1964).
26. McCracken, R. A. "Study of Color and Iron Removal by Means of Pilot Plant at Amesberry," *New Engl. Water Works Assoc.* **75**, 102-114 (1961).
27. Kollin, W. "Iron Removal Methods By Diatomite Filtration," *Water Sewage Works* **109**, R182 (1962).
28. Shireman, H. C. "Filtration Boosts Tertiary Treatment," *Water Wastes Eng.* **9**, 34-37 (April, 1972).
29. Nebolsine, R. and R. Sanday. "Ultra-High Rate Filtration, New Technique for Purification and Reuse of Water," *Iron Steel Eng.* **XLIV**, 105-112 (December, 1967).
30. *The Cost of Clean Water, Vol. III, Industrial Waste Profiles, No. 1, Blast Furnace and Steel Mills,* FWPCA Publication (Washington, D.C.: U.S. Dept. of the Interior, 1967).
31. "Water Pollutant or Reusable Resources?" *Environ. Sci. Technol.* **4**, 380-382 (1970).
32. Heise, L. W. and M. Johnson. "Practical Development Aspects of Waste Pickle Liquor Disposal," *Proc. 13the Purdue Ind. Waste Conf.* **13**, 140-150 (1958).
33. Melzer, S. F. and T. L. Taubken. "New Process Treats Acid Rinse Waters, *Water Wastes Eng.* **8**, F6-F8 (November, 1971).
34. Cupps, C. C. "Treatment of Wastes for Automobile Bumper Finishing," *Ind. Water Wastes* **6**, 111-114 (1961).
35. Spencer, E. F. Jr. "Pollution Control in the Chemical Industry," In *Industrial Pollution Control Handbook,* Herbert F. Lund, Ed. (New York: McGraw-Hill Book Co., 1971).
36. Delaine, J. "Management to Achieve Water Economy in the Iron and Steel Industry—Conclusion," *Effl. Water Treat. J.* **6**, 543-551 (1966).
37. Besselievre, E. B. *The Treatment of Industrial Wastes* (New York: McGraw-Hill Book Co., 1969).
38. Jackson, D. V. "Metal Recovery from Effluents and Sludges," *Metal Finishing J.* **18**, 235-242 (1972).

12

TREATMENT TECHNOLOGY FOR LEAD

INDUSTRIAL SOURCES

Lead is used as an industrial raw material for storage battery manufacture, printing, pigments, fuels, photographic materials, and matches and explosives manufacturing. It is one of the most widely used nonferrous metals in industry.

The storage battery industry is the largest consumer of lead, followed by the petroleum industry in producing gasoline additives (1). In a survey of eight battery manufacturing plants, it was reported that lead losses per battery manufactured ranged from 4.54–6810 mg, and that water usage ranged from 11–77 gal per battery (2). Highest lead concentrations originated from the plate-forming area. This water has a low pH, which leads to a high concentration of dissolved lead. Waste pH values of 1.3–2.6 have been reported (3). Particulate lead in battery wastes is very fine, and is reported to require long detention periods (above 24 hr) for effective settling (2).

The use of lead (as sodium plumbite) in petroleum refining produces a lead sludge, from which lead is normally recovered (4). Lead is also a common constituent of plating wastes, although not so frequently encountered as copper, zinc, cadmium and chromium. The lead content of wastewater from one engine parts plating plant has been reported as ranging from 2.0–140.0 mg/l (5). The high concentrations resulted from dumping of spent plating bath solution. Lead content of plating bath rinse waters has been reported to occur at up to 30 mg/l of lead (6).

A lead solder-glass frit mixture is used to fuse the front glass panel to the funnel of television tubes. Quality control of color picture tubes may require the salvage and reuse of the glass envelopes. Separation of the two components is accomplished by dissolving the frit in dilute nitric acid, which yields an acidic wastewater containing lead and fluoride. Wastewater lead levels of 400 mg/l have been reported from such a salvage operation (7).

Pollution from lead mines, mining, and smelting has also been reported in the literature (8-11). Analyses of lead mine wastewaters are presented in Table 42. Initial lead levels shown represent wastewater from the milling process, and final values are those found following long-term detention of the wastewater in holding ponds. Pond detention times were not reported. Initial levels ranging from 0.018–0.098 mg/l and effluent levels of 0.022–0.055 mg/l were found (12). With one exception, lead reductions of approximately 45% were observed. This is likely due primarily to sedimentation of the particulate lead contained in the wastewater.

Table 42. Lead Mine Milling Process Wastewater (12).

Lead Concentration (mg/l)		
Initial	Final[a]	Percent Removal
0.040	0.022	45
0.018–0.036	0.038	+
0.051	0.030	44
0.098	0.055	44

[a] Effluent from holding pond.

Municipal water pumped from an abandoned mine shaft in Australia, which was used as a municipal water supply infiltration gallery, was reported to contain 0.31 mg/l of lead (10). Significant levels of lead have also been found in acid mine drainage, as shown in Table 43. At the reported pH values, the lead would be predominantly in soluble form.

Most lead wastes are in the inorganic particulate or ionic form. However, wastewater of the tetraethyl lead industry contains sig-

Table 43. Lead Content of Acid Mine Drainage (13).

Source	pH	Lead (mg/l)
Tailing Pond	2.0	0.67
Mine Drainage	2.6	0.3
Mine Drainage	3.2	2.5
Mine Drainage	—	0.02

nificant concentrations of organic lead compounds. Because these compounds do not respond to the standard precipitation treatment procedures employed for inorganic lead, wastewaters of the tetraethyl lead industry present particular difficulty. Table 44 summarizes reported lead levels in industrial wastewaters.

Table 44. Reported Lead Levels in Industrial Wastewaters.

Industry	Lead (mg/l)	Reference
Battery Manufacture		2
Particulate	5–48	
Soluble	0.5–25	
Battery Manufacture		3
Particulate	0.4–66.5	
Soluble	2.6–5.1	
Plating	2–140	5
Plating	0–30	6
Television Tube Manufacture	400	7
Mine Drainage	0.02–2.5	13
Mine Drainage	0.04–0.5	11
Mining Process Water	0.018–0.098	12
Lead Smelting	0.07–0.16	14
Ammunition Plant	6.5	15
Tetraethyl Lead Manufacture		16
Organic	126.7–144.8	
Inorganic	66.1–84.9	
Tetraethyl Lead Manufacture	45	17

TREATMENT TECHNOLOGY

Treatment of soluble lead involves, for the most part, reaction to form a lead precipitate, and sedimentation. Precipitation chemicals used include lime or caustic to precipitate lead hydroxide, soda-ash to precipitate lead carbonate, and phosphate to precipitate lead phosphate. Alum, and ferrous and ferric sulfate coagulation have also been reported for lead treatment, as has ion exchange.

Precipitation

In the precipitation processes, lead is normally precipitated as the carbonate, $PbCO_3$ or the hydroxide, $Pb(OH)_2$. The literature contains conflicting values of pH optima for precipitation of lead hydroxide. For example, a lead precipitation pH of 6.0, has been suggested in one instance (18), and a pH range of 6.0–7.0 has been suggested in another (13). Effective lead treatment of a metal finishing waste has also been reported at pH 8.7–9.3 (19), and even higher pH (up to 10.0) has also been recommended (20,21). Bench-scale studies on the effect of pH on lead hydroxide precipitation indicate an optimum pH of 10.0 yielded a residual lead level below 0.5 mg/l (20). Both theoretical solubility calculations (22) and plant data (19,23) indicate that most efficient lead hydroxide precipitation occurs at pH 9.2–9.5. Effluent lead of 0.01–0.03 has been reported for treatment in this pH range (23). Efficiency of hydroxide precipitate formation decreases rapidly at more extreme pH values. Lead is often reported as being recovered from the sludge formed in the precipitation processes (4,24).

In forming insoluble lead hydroxide, lime is the treatment chemical of choice, although caustic has been used. Lead in a municipal drinking water supply has been reported reduced by lime treatment and settling from an initial value of 0.31 mg/l down to 0.1 mg/l or less of lead (10).

A survey of lead wastewater treatment practices at eight battery manufacturing plants indicated that simple sedimentation was relatively ineffective in removing particulate lead and, of course, ineffective in reducing dissolved lead (2). Data reported for four plants that used pH adjustment, followed by sedimentation (for three plants) are presented in Table 45. Improved effluent levels were observed at more alkaline pH and for long settling times. The benefit of effective settling is demonstrated by Plant D, which had

Table 45. Treatment Practice and Effluent Lead Levels from Battery Plants (2).

Plant	Treatment Chemical	pH	Settling Time (hr)	Effluent Lead (mg/l)		
				Dissolved	Suspended	Total
A	Lime	6.8	30	0.30	0.18	0.48
B	Caustic	5.5	1	0.6	1.0	1.6
C	Caustic	7.0	24	0.02	0.02	0.04
D	Ammonia[a]	7.8	0	1.9	22	23.9

[a] Used for pH adjustment.

no settling and a poor effluent. Plant D also had the poorest soluble lead effluent, probably as a result of using a complexing agent (ammonium hydroxide) for pH adjustment. Complexation of lead with ammonia would prevent effective precipitation. Best effluent, among the plants surveyed, was obtained by addition of caustic (NaOH) to pH 7.0, followed by 24-hour settling.

Solubility of lead carbonate is dependent upon both treatment pH and carbonate concentraton. For a fixed carbonate dosage, a broad range of pH (8.0–9.2) provides consistent treatment (22,23). In pilot studies on carbonate precipitation of lead battery wastes, sodium carbonate was employed as the treatment chemical (3). Precipitation plus sand filtration yielded effluent total lead of 0.2–3.8 mg/l, and soluble lead of 0.1 mg/l. Treatment was at pH 6.4–8.7.

Pilot plant work on lead wastes emphasizes the importance of effective solid removal. The results of this work on 4-gpm continuous flow pilot plants are shown in Table 46. Initial lead concentration was 5 mg/l, and filtration was through dual media. As the data of Table 46 reveal, effluent levels almost an order of magnitude lower were achieved by sedimentation plus filtration, as opposed to sedimentation only.

Precipitation of dissolved lead by dolomite ($CaCO_3 \cdot MgCO_3$) to yield lead carbonate has been reported (26). The process involved filtration of lead wastewater through a calcined dolomite bed, which reacted with and precipitated the dissolved lead. The dolomite filters were regenerated by backwashing with water, which removed the lead carbonate formed in the process.

The reaction of lead with dolomite, yielding insoluble lead carbonate, results in very little lead entering receiving streams from southeastern Missouri lead mines (8,9,12). The lead veins are located in a dolomite deposit, and the mine tailings thus contain crushed dolomite. As the mine tailings are channeled through sedimentation basins prior to release of the waste effluent, dolomite plus

Table 46. Lead Removal by Precipitation and Sedimentation plus Filtration (25).

Treatment Chemical	Treatment pH	Effluent Lead Concentration (mg/l)	
		Sedimentation	Filtration
Ferric Sulfate at 45 mg/l Fe	6.0	0.25	0.030
Lime at 260 mg/l + Ferric Sulfate at 20 mg/l Fe	10.0	0.25	0.029
Lime at 600 mg/l	11.5	0.20	0.019

reacted lead carbonate are readily removed. Under existing conditions, any heavy metals in the wastewater are precipitated rapidly in the settling lagoons at the slightly basic pH (7.5–8.2) of the waste.

The use of sodium phosphate (Na_3PO_4) has been reported for lead treatment, to precipitate lead phosphate (3,7). The solubility of lead phosphate is determined by treatment pH and phosphate dosage. Treatment conditions generally specified are pH above 7.0, with up to three times excess sodium phosphate added (3.7 lb sodium phosphate/lb lead). This may however result in unacceptable effluent phosphate levels. Treatment of a battery waste with sodium phosphate at pH 7.2–7.4 has been reported to achieve 0.2–0.6 mg/l total lead and 0.1 mg/l soluble lead (3). Equivalent lead treatment has been reported for phosphate treatment of an electronic tube manufacturing waste (7).

Coagulation

Treatment of a wastewater from a tetraethyl lead plant has been described (24). The effluent was alkaline, and contained inorganic lead salts in solution and suspension, plus organic lead salts. The waste pH was adjusted to 8–9, and inorganic lead precipitated by addition of lime. The precipitation was assisted by the addition of ferrous sulfate, which acted as a coagulating agent. The treated effluent flowed to a sedimentation basin, from which the lead sludge was collected, dewatered in a filter, and reclaimed by a lead slag refinery.

Nozaki and Hatotani (16) also have described treatment of wastewaters from a tetraethyl lead manufacturing process. The two major categories of waste were inorganic lead wastewaters and organic lead wastewaters. After sedimentation in a holding basin to recover solid lead and lead oxide, the inorganic lead waste fraction (66.1 mg/l) was effectively treated by coagulation with ferric and ferrous sulfate. Treatment of combined organic and inorganic lead waste streams containing respective concentrations of 13.3 mg/l and 84.9 mg/l, was not as effective, however. Effluent contained "trace" to 1.4 mg/l of inorganic lead, and most of the influent organic lead. A separate bench-scale treatment process was therefore developed for the organic lead waste stream. The organic lead compounds were removed effectively on a strongly acid cation exchange resin. Reductions of organic lead from 126.7–144.8 mg/l to 0.020–0.53 mg/l (as lead) were reported. The organic lead compounds adsorbed on the ion exchange resin were eluted by caustic soda. This eluate was collected in a reaction basin, into which chlorine gas was injected at 95°C for 45 minutes. The organic lead compounds were thereby almost completely converted to inorganic lead compounds, which could be treated as described above. Overall lead removals to less than 1 mg/l residual were expected for the final plant effluent, after installation of the organic lead wastewater treatment sequence (16).

A third report on treatment of lead-bearing waste from a tetraethyl plant reported a similar treatment sequence for inorganic lead (17). Lead occurring in the manufacturing process effluent at a concentration of 45 mg/l was reduced to 1.7 mg/l by precipitation at pH 10.4–10.8, aided by addition of ferrous sulfate. The lead sludge was recovered, and lead was extracted and reclaimed. The author reports that $500,000 (1960 prices) was spent on waste treatment facilities, out of a total plant investment of $12 million. Waste treatment capital costs thus represented 4.2% of the total plant costs. Waste treatment included settleable solids and chloride control, in addition to lead removal.

Similar treatment has been reported for lead removal from a municipal water supply (27). Lime and ferrous sulfate are added to the water to precipitate and coagulate lead leached from a nearby abandoned mine. Alum coagulation at pH 6.8–7.0 has been reported to reduce lead from 17 to 1.3 mg/l (21).

Ion Exchange

Successful ion exchange treatment has been reported for both organic and inorganic lead (15,16,28). In one study on treatment of an ammunition industry waste, lead was initially reduced from 6.5 to 0.1 mg/l by precipitation. Ion exchange treatment of the effluent at pH 5.0–5.2 with a phosphoric acid resin reduced the lead to 0.01 mg/l (15). In another application, an effluent level of 0.002 mg/l was achieved (28). Organic lead reductions for 126.7–144.8 mg/l to 0.02–0.53 mg/l were reported, using a strong acid cation exchange resin (16).

SUMMARY

Information on the treatment of lead-bearing wastewaters is summarized in Table 47. Precipitation by lime or dolomite and sedimentation, with sludge disposal or lead recovery, have been shown to be effective, as has ion exchange.

Treatment efficiencies by precipitation plus sedimentation ex-

Table 47. Summary of Reported Lead Treatment Processes and Efficiencies.

Treatment	pH	Lead Concentration (mg/l) Initial	Final	Percent Removal	Reference
Ion Exchange	5.0–5.2	0.1	0.01	90	15
Ion Exchange[a]		0.055	0.0015	97.3	28
Ion Exchange[a]		126.7–144.8	0.020–0.53	99.9+	16
Lime + Sedimentation		0.31	<0.1	97+	10
Lime + 30-hr Sedimentation	6.8		0.48		2
Caustic + 1-hr Sedimentation	5.5		1.6		2
Caustic + 24-hr Sedimentation	7.0		0.04		2
Ammonium Hydroxide	7.8		23.9		2
Lime + Ferric Sulfate + Sedimentation[a]	10.0	5.0	0.25	95	25
+ Filtration[a]		5.0	0.029	99.4	
Lime + Sedimentation[a]	11.5	5.0	0.20	96	25
+ Filtration[a]		5.0	0.019	99.6	
Lime + Sedimentation		6.5	0.1	98.5	15
Lime[a]	9.0–9.5	5.0	0.01–0.03	99+	23
Caustic + Soda Ash[a]	9.0–9.5	5.0	0.01–0.03	99+	23
Dolomite + Sedimentation	7.5–8.2	0.04	0.022	45	12
		0.018–0.036	0.038	+	
		0.051	0.030	42	
		0.098	0.055	44	
Sodium Carbonate + Filtration[a]	6.4–8.7	10.2–70.0	0.2–3.6	82–99+	3
Sodium Phosphate + Filtration[a]	7.2–7.5	3.0–5.0	0.2–0.6	83–93	3
Ferric Sulfate + Sedimentation[a]	6.0	5.0	0.25	95	25
+ Filtration[a]		5.0	0.03	99.4	
Ferrous Sulfate + Sedimentation	10.4–10.8	45	1.7	96.2	17

[a] Pilot Plant Results

ceeding 99% have been observed in both pilot plant and full-scale operations, and pilot plant results on a 240 gal/hr unit indicate that filtration can improve effluent levels an order of magnitude. There is also an indication that long settling tank detention times (one day or greater) may match filter performance.

REFERENCES

1. Hall, S. K. "Lead Pollution and Poisoning," Environ. Sci. Technol. **6**, 30-35 (1972).
2. Haas, W. R. and S. Miller. "Evaluation of Lead Plant Wastewater Treatment Methods," IIT Research Institute Report No. IITRI-C8213-2. (1972).
3. Crandall, C. J. and J. R. Rodenberg. "Waste Lead Oxide Treatment of Lead Acid Battery Manufacturing Wastewater," presented at 29th Ind. Waste Conf., Purdue University (May 7, 1974).
4. Hill, J. B. "Waste Problems in the Petroleum Industry," *Ind. Eng. Chem.* **31** (11), 1361-1363 (1939).
5. Fales, A. L. "A Plating Waste Disposal Problem," *Sewage Works J.* **20**, 857-860 (1948).
6. Pinkerton, H. L. "Waste Disposal," In *Electroplating Engineering Handbook*, 2nd ed., A. Kenneth Graham, Editor-in-chief (New York: Van Nostrand Reinhold Co., 1962).
7. Rohrer, K. L. "An Integrated Facility for the Treatment of Lead and Fluoride Wastes," *Ind. Water* **17** (9), 36-39 (1971).
8. Wixson, B. G., E. A. Bolter, N. H. Tibbs and A. R. Handler. "Pollution from Mines in the New Lead Belt of Southeastern Missouri," *Proc. 24th Purdue Ind. Waste Conf.* 632-643 (1969).
9. Jennett, J. C. and B. G. Wixson. "Treatment and Control of Lead Mining in S.E. Missouri," *Proc. 26th Purdue Ind. Waste Conf.* (1971).
10. Dept. of Mines, Australia. "Annual Report Western Australian Government (1962).
11. Larsen, H. P., J. K. P. Shou and L. W. Ross. "Chemical Treatment of Metal-Bearing Mine Drainage," *J. Water Poll. Cont. Fed.* **45** (8), 1682-1695 (1973).
12. Bolter, E., J. C. Jennett and B. G. Wixson. "Geochemical Impact of Lead-Mining Wastewaters on Streams in Southwestern Missouri," Presented at 27th Ind. Waste Conf., Purdue University, 1972.
13. Mill, R. D. "Control and Prevention of Mine Drainage," Presented at Conf. on Recycling and Control of Metals, Battelle-Columbus, 1972.
14. Hallowell, J. B., J. F. Shea, G. R. Smithson, Jr., A. B. Tripler and B. W. Gonser. "Water Pollution Control in the Primary NonFerrous Metals Industry Volume I. Copper, Zinc and Lead Industries," U.S. Environmental Protection Agency, Report EPA-R2-73-247a (September, 1973).

15. Shannon, R. E. "Some Examples of the Concentration of Trace Heavy Metals with Ion Exchange Resins," In *Traces of Heavy Metals in Water, Removal Processes and Monitoring, Proc. Center for Environmental Studies,* Princeton University (1973).
16. Nozaki, M. and H. Hatotani. "Treatment of Tetraethyl Lead Manufacturing Wastes," *Water Res.* **1** (2), 167-177 (1967).
17. Robb, L. A. "Waste Water Treating Facilities at Ethyl Corporation of Canada, Ltd.," *Proc. 8th Ontario Ind. Waste Conf.* 90-96 (1961).
18. Dean, J. G., F. L. Bosqui and K. L. Canouette. "Removing Heavy Metals from Waste Water, *Environ. Sci. Technol.* **6,** 518-522 (1972).
19. "Effluent Treatment at B.S.R.," *Metal Finishing J.* **17,** 201-248 (1971).
20. Netzer, A., A. Rowers and J. D. Norman. "Removal of Trace Metals from Wastewater by Lime and Ozonation," In *Pollution: Engineering and Scientific Solutions* (in press).
21. Nilsson, R. "Removal of Metals by Chemical Treatment of Municipal Waste Water," *Water Res.* **5,** 51-60 (1971).
22. Hem, J. D. and W. H. Durum. "Solubility and Occurrence of Lead in Surface Water," *J. Amer. Water Works Assoc.* **651,** 562-568 (1973).
23. Day, R. V., E. T. Lee and E. S. Hochuli. "Bell Systems' Metals Recovery Plant," *Ind. Wastes* **20** (4), 26-29 (1974).
24. "An Integrated Plant for Tetraethyl Lead," *Ind. Chem.* **30,** 429-436 (1954).
25. Maruyama, T., S. A. Hannah and J. M. Cohen. "Removal of Heavy Metals by Physical and Chemical Treatment Processes," Presented at 45th Water Poll. Cont. Fed. Conf., 1972.
26. Voznesenskic, S. A., A. B. Evalonova and R. V. Suvorova. "Physico-Chemical Methods of Purification of Industrial Waste Waters," *Water Poll. Abstr.* **13,** 135 (1940).
27. "East Midland Districting Meeting," *J. Inst. Municip. Eng.* **84,** XV-XXVCC (November, 1957).
28. Liebig, C. F., Jr., A. P. Vanselow and H. D. Chapman. "The Suitability of Water Purified by Synthetic Ion-Exchange Resins for the Growing of Plants in Controlled Nutrient Cultures," *Soil Sci.* **55,** 371-376 (1943).

13

TREATMENT TECHNOLOGY FOR MANGANESE

INDUSTRIAL SOURCES

Manganese and its salts are used in the following industries (1):

- Steel alloy
- Dry cell battery
- Glass and ceramics
- Paint and varnish
- Ink and dye
- Fertilizers
- Match and fireworks

Among the many forms and compounds of manganese, only the manganous salts and the highly oxidized permanganate anion are appreciably soluble. The latter is a strong oxidant, which is reduced under normal circumstances to insoluble manganese dioxide. The highly soluble manganous chloride is used in dyeing operations, linseed oil driers and electric batteries, while the equally soluble manganous sulfate is used in porcelain glazing, varnishes and production of specialized fertilizers.

Manganese concentrations of 1.2–8.0 mg/l have been reported for mine waters (2), with a value of 50 mg/l for an acid mine water (3).

Since the chemistry of manganese is similar to that of iron (4), it would be expected that any pickling operation involving manganese steel alloy results in dissolution of manganous ion and the presence of this ion in pickling and rinse solutions. By contrast to iron, the divalent (manganous) form is not readily oxidized to the insoluble manganic form, other than at elevated pH (4).

TREATMENT TECHNOLOGY

Removal of soluble manganese has been practiced for many years in the treatment of municipal water supplies and industrial process waters. Low manganese waters are essential for many industries, including food and beverage. Treatment technology basically involves conversion of the soluble manganous ion to an insoluble precipitant. In water treatment, this generally means simultaneous removal of iron, since conditions under which high iron levels occur in water supplies are essentially the same as those for soluble manganese. Removal is effected by oxidation of the manganous ion, and by separation of the resulting insoluble oxides and hydroxides of manganese.

Manganese removal processes are listed in Table 48. Because of the low reactivity of manganous ion with oxygen, simple aeration is not an effective oxidation technique below pH 9, and even above that pH the oxidation reaction is slow (5). Organic matter in solution can combine with manganese and prevent its oxidation by simple aeration, even at high pH (6). Figure 30 illustrates the effect of pH on manganese removal by lime. A reaction pH above 9.4 is required to achieve significant manganese reduction.

Table 48. Manganese Removal Processes (5).

Process	Comments
1. Aeration	Slow and ineffective below pH 9.
2. Chlorine and Hypochlorite Oxidation	Not effective for organically bound manganese.
3. Adjustment of pH	Lime-soda type treatment gives removal at pH 9.5
4. Catalysis	Copper ion enhances air oxidation.
5. Ion exchange	Effective for small quantities of iron and manganese. Resins quickly fouled by iron and manganese oxides.
6. Chlorine Dioxide	Rapidly oxidizes manganese to the insoluble form, but expensive.
7. Manganese Dioxide or Potassium Permanganate	Regeneration, greensand filter process. Economic disadvantage of requiring excess permanganate.
8. Direct Potassium Permanganate Addition	Requires filtration (sand, diatomite, or mixed media).

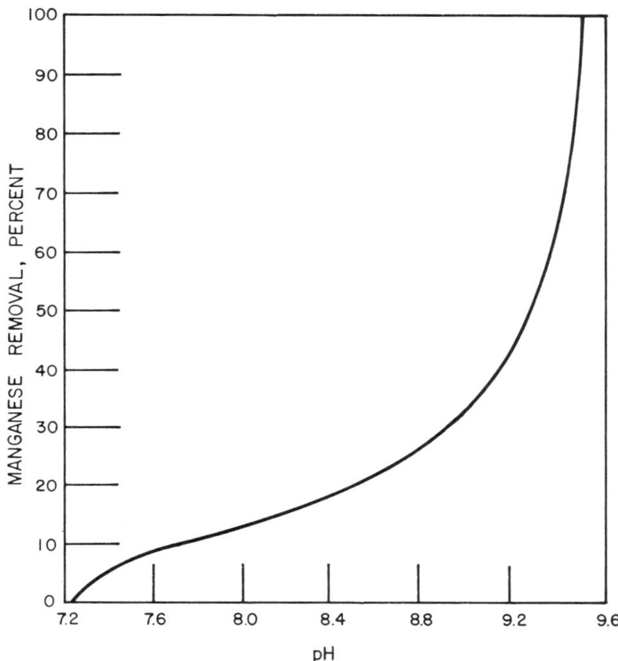

Figure 30. Effect of pH on efficiency of manganese removal using lime. [From Owens (7), courtesy of *Water and Sewage Works.*]

Hill (2) indicates that manganese can be reduced to less than 0.5 mg/l by simple neutralization and precipitation, but no pH values are reported. He states further that the use of calcium carbonate plus lime for iron and manganese removal in Japanese acid mine waters has proved successful.

The use of chemical oxidants to convert manganous ion to insoluble manganese dioxide, in conjunction with coagulation and filtration, has been recommended, although it may cause filtration problems. The character of manganese flocs produced when strong oxidants are used has been reported to rapidly clog slow sand filters (6). Despite reports that manganese oxidation by chlorination may not be effective (5), successful manganese removals have been reported. Owens (7) reports removal of manganese at concentrations as high as 5.7 mg/l, with chlorination followed by lime coagulation and rapid sand filtration. Split flow treatment was employed to achieve intermediate manganese levels in the final effluent. Chemical costs, adjusted to the treated stream only, were about 2.1¢/1000 gal. Sodium silicate was used as a coagulant aid. Effective manganese oxidation by chlorination requires a free chlorine residual of 0.5 mg/l and a dosage of 1.29 mg of chlorine per mg of manganese oxidized (8).

The successful use of potassium permanganate as a chemical oxidant has been reported. Addition of 1.5–2.0 mg/l of potassium permanganate reduced 1.0 mg/l of manganese to less than 0.05 mg/l (9). Addition is 30 minutes upstream from the prechlorination point, which is followed by lime-alum-sodium silicate coagulation (4–10 hours detention), and rapid sand filtration. Costs attributed to permanganate addition are less than 0.3¢/1000 gal. Since iron is generally present, the quantity of permanganate used must include that required for oxidation of the iron content. The stoichiometric relationship between permanganate required and the quantity of iron and manganese present is shown in Figures 31 and 32 respectively. The ability to remove manganese at lower pH (an advantage when simultaneous iron removal is desired) by permanganate oxidation is demonstrated by the data of Holland (11). The addition of 0.04 mg/l of sodium alginate reduced 0.35 mg/l of manganese to less than 0.02 mg/l in the filtered supernatant. The absence of residual taste or odor has been cited as an additional advantage of permanganate over chlorination (6).

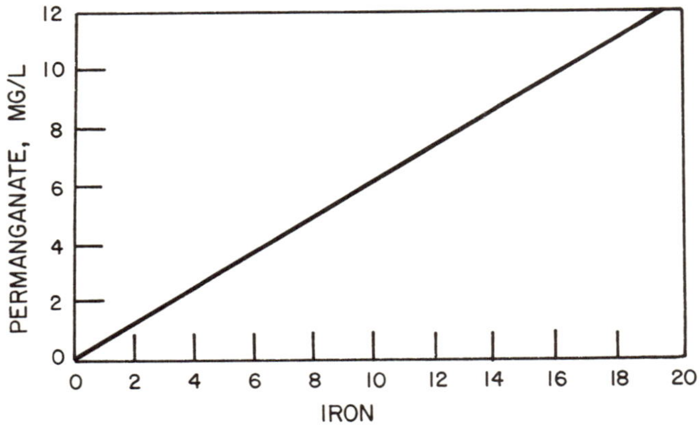

Figure 31. Permanganate requirement for oxidation of ferrous iron. [From Willey and Jennings (10), courtesy of American Water Works Association.]

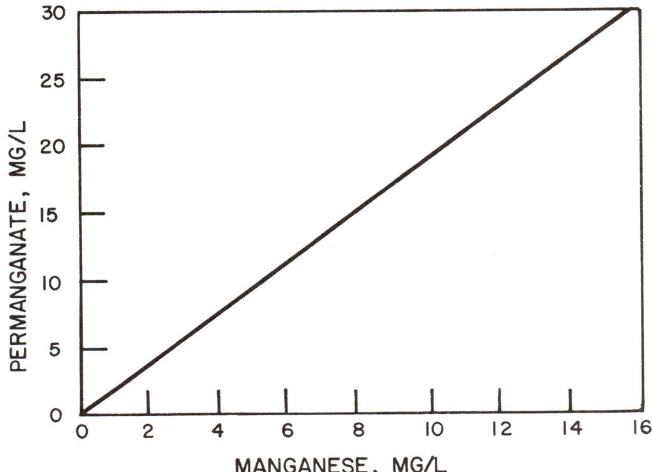

Figure 32. Permanganate requirement for oxidation of manganous manganese. [From Willey and Jennings (10), courtesy of American Water Works Association.]

The application of permanganate oxidation in conjunction with vacuum diatomite filtration has been used to treat a municipal water supply (12). Manganese is reduced from 1.23 mg/l to 0.05 mg/l or less. Costs for the 2.33-MGD unit were given as $107/1000 gpd. No operating costs were provided but the unit requires a total of 250 pounds of diatomite per million gallons of water filtered.

Permanganate may be used indirectly for manganese removal in the manganese greensand filter method. The filter is a combination of New Jersey Glauconite (greensand), manganese sulfate and potassium sulfate, which results in a filter bed of sand grains containing high oxides of manganese (10). Upon contact with manganous ion an oxidation-reduction reaction occurs, resulting in the formation of insoluble manganese dioxide, which remains in the filter bed. The bed is periodically reoxidized with permanganate, and solids are backwashed from the bed. With this system, manganese was reduced from 1.7 mg/l to 0.0 mg/l at pH 7.2 using a pressure greensand system (13). The plant, of 500–700 gpm capacity, cost $26/1000 gpd completely installed and operating, compared with an estimated $74/1000 gpd for a conventional aeration, lime coagulation-clarifier-filter-recarbonation system. Chemical costs were 3.5¢/1000 gal for a permanganate dosage of 13 mg/l. The water had a manganese concentration of 13 mg/l, which required excess permanganate. Treatment costs were roughly equivalent to lime chemical costs required for a similar treatment level. Regeneration for this plant was on a continuous basis with periodic backwashing of the greensand filter. Generalized removal capacities for greensand filters are given as 0.09 pounds of manganese per cubic foot of filter volume with regeneration requirements of 0.18 lb/ft^3 of potassium permanganate. Flow rates of 3 $gal/min/ft^2$ are normal (14).

The use of ozone in conjunction with lime has been investigated as a manganese removal technique in laboratory studies. The data of Table 49 presents results for solutions of manganous chloride initially containing 94 mg/l manganese.

Table 49. Ozone Treatment for Manganese (15).

	Final Manganese Conc. (mg/l)	
pH	Lime Only	Lime Followed by Ozone
7.0	—	0.8
8.0	90	0.3
9.0	17	<0.1
10.0	0.15	<0.1
11.0	<0.1	.—

Ion exchange has been utilized for iron and manganese removal. Many advantages are cited, including smaller capital investment than coagulation-filtration, nearly four times greater flow rates, smaller plant requirements and simple operation (14). The obvious drawback to ion exchange is the nonselective removal of other ions, which increases operation costs. Regeneration brine disposal poses an additional treatment problem, since the manganese remains in soluble form in the brine. An anaerobic waste stream is necessary for the ion exchange system, to avoid air oxidation of the manganese with resultant precipitation and clogging of the ion exchange bed. Treatment data for one ion exchange unit over three years of operation are summarized in Table 50.

The application of the above processes to manganese removal from wastewaters should be direct, within the constraint that the presence of acids, other oxidizable metals, or organics in the waste stream may substantially increase costs related to manganese removal. The unit operation costs for coagulation-sedimentation-filtration should be similar to those for iron treatment systems.

Capital and operating costs have been reported for mixed media

Table 50. Efficiency of Ion-Exchange Treatment of Water (14).

Constituent	Raw Well Water Concentration (mg/l)	Treated Water Concentration (mg/l)
	Eight Months of Operation	
Hardness	96	2
Sodium	49	224
Potassium	6.5	3
Bicarbonate	62	130
Free CO_2	95	50
Total Iron	37	0.5
Total Manganese	1.7	0.1
pH	6.1	6.7
	Fifteen Months of Operation	
Hardness	86	2
Bicarbonate	50	138
Total Iron	53	0.3
Total Manganese	1.0	0.0
pH	5.7	6.4
	Thirty-Six Months of Operation	
Hardness	76	2
Sodium	74	280
Bicarbonate	15	132
Free CO_2	80	50
Total Iron	52	0.1
Total Manganese	1.3	0.0
pH	5.6	6.5

filtration in advanced municipal waste treatment (16). Capital costs were \$65/1000 gpd and operating costs were 2.3¢/1000 gal based on a 7.5 MGD effluent. The 7.5-MGD unit actually filters 9.3 MGD when recycle volumes are included. Thus operating costs are 1.8¢/1000 gal based on the total volume through the filter. Additional costs would be attributable to chlorine (or chlorine dioxide) and chlorination equipment, plus lime necessary to meet the higher pH requirements. Chemical costs could be somewhat lower or higher, depending upon the nature of the industrial waste to be treated.

SUMMARY

Manganese removal to levels of 0.05 mg/l or less have been accomplished for many years in the treatment of municipal drinking and industrial process waters. Several methods of removal are available, each of which may have distinct advantages depending upon the overall nature of the waste, other required waste treatment operations, and the total treatment goals for a particular waste. The most common general approach seems to involve oxidation of the soluble manganous form to insoluble manganese hydroxide or oxide, with subsequent precipitation. Ion exchange treatment has proven effective, however, and has several advantages.

REFERENCES

1. McKee, J. E. and H. W. Wolf. *Water Quality Criteria,* California State Water Quality Control Board Publication No. 3-A (1963).
2. Hill, R. D. "Control and Prevention of Mine Drainage," Presented at the Environmental Resources Conference on Cycling and Control of Metals, Battelle Columbus Laboratories, Columbus, Ohio, 1972.
3. Kunin, R. and D. G. Downing. "Ion Exchange System Boasts More Pulling Power," *Chem. Eng.* **78**, 67-69 (June 28, 1971).
4. Cotton, F. A. and G. Wilkinson. *Advanced Inorganic Cehmistry* (New York: Wiley-Interscience, 1962).
5. Welch, A. "Potassium Permanganate in Water Treatment," *J. Water Works Assoc.* **55**, 734-741 (1963).
6. Borgren, G. R. "Removal of Iron and Manganese from Ground Waters," *J. New Engl. Water Works Assoc.* **76**, 70-76 (1962).
7. Owens, L. V. "Iron and Manganese Removal by Split Flow Treatment," *Water Sewage Works* **110**, R76-R87 (1963).
8. Harris, W. C. "Iron and Manganese Removal," *Eng. Exp. Sta. Bull.* No. 68, Louisiana State University (1962).
9. Haney, C. G. "Potassium Permanganate Solves a Problem of Charlottesville, Va.," *Water Sewage Works* **111**, R353-R354 (1964).
10. Willey, B. F. and H. Jennings. "Iron and Manganese Removal with Potassium Permanganate," *J. Amer. Water Works Assoc.* **55**, 729-734 (1963).
11. Holland G. J. "Removal of Dissolved Aluminum, Iron and Manganese," *Chem Ind. (London)* 2016 (1964).
12. Costabile, F. J. and C. H. Perron. "Diatomite Filter Ends Manganese Problem," *J. Amer. Water Works Assoc.* **63**, 230-232 (1971).
13. Alling, S. F. "Continuously Regenerated Greensand for Iron and Manganese Removal," *J. Amer. Water Works Assoc.* **55**, 749-752 (1963).
14. Alzentzer, H. A. "Ion Exchange in Water Treatment," *J. Amer. Water Works Assoc.* **55**, 742-748.
15. Netzer, A., A. Bowers and J. D. Norman. "Removal of Trace Metals from Wastewater by Lime and Ozonation," In *Pollution: Engineering and Scientific Solutions* (in (press).
16. Shireman, H. C. "Filtration Boosts Tertiary Treatment," *Water Wastes Eng.* **9**, 34-37 (April, 1972).

14

TREATMENT TECHNOLOGY FOR MERCURY

INDUSTRIAL SOURCES

The major consumptive user of mercury, Hg, is the chloralkali industry, using the DeNora or similar electrolytic cells (1-6), whose losses to waste discharge without process modification or wastewater treatment have been variously reported as 0.1–0.5 lb Hg/ton of chlorine produced (2,3,5). The second largest user is the electrical and electronics industry (2,3,5). Other users of various mercury forms include: explosives manufacturing, the photographic industry, and the pesticide and preservative industry (1,2,4-6).

Within the chemical and petrochemical industry, mercury compounds have found use as catalysts in plastics production, hydrogenation and dehydrogenation, sulfonation, oxidation, chlorination and acidolysis (1,4).

Both the paint and pharmaceutical-cosmetic industries have been users of mercury compounds as preservatives and antifouling or mildew-proofing agents (1,2,4). Pulp and paper operations have used mercurials as fungicides, slimicides and biocides (1,2,4), but recently use has been curtailed (2). Other sources of mercury wastes are seed preservatives, research and hospital labs, sealants

Table 51. Mercury Consumed in the United States in Tons per Year (2).

Use	Year 1965	1966	1967	1968	1969
Agriculture	118	90	142	130	102
Amalgamation	10	10	8	10	7
Catalysts	35	73	102	73	112
Dental Preparations	62	51	52	79	117
Electrical Apparatus	612	618	555	664	709
Electrolytical Cl_2 and NaOH Production	333	438	544	663	787
General Laboratory Use	42	59	43	47	78
Industrial and Control Instruments	179	157	147	149	265
Paint—antifouling	10	5	6	15	9
—mildew-proofing	312	315	267	387	360
Paper and Pulp	23	23	17	16	22
Pharmaceuticals	17	9	11	16	28
Other	585	594	478	302	368

Table 52. Levels of Mercury in Industrial Effluents (9).

Waste	Dissolved Mercury (mg/l)	Suspended Solids Mercury (mg/kg) dry weight of solids
Paper Mill Settling Pond	.00008	10
Paper Mill Effluent	.002–.0034	5.6
Fertilizer Mill	.00026–.004	32.0
Smelting Plant	.002–.004	—
Chloralkali Plant	.080–2.0	14.0

used in machines, thermometers and manometers, and film production and processing (1,4).

A summary of consumptive use of mercury in the United States from 1965–1969 is presented in Table 51. It has been pointed out in conjunction with these data that the total number of direct and indirect users of mercury extends into the millions (2).

Limited data are available regarding actual wastewater mercury levels. Wastewater emanating from an acetaldehyde production unit was reported to contain 20 mg/l of mercury and the combined wastewater, prior to treatment, contained mercury at 18 mg/l (7). Effluent mercury from one chloralkali plant is estimated to have been 1.4–2.8 mg/l, based on 600-gpm waste discharge and 10–20 lb/day mercury loss (8). Additional mercury effluent data are shown in Table 52.

It has been reported that commercial bleach preparations alone contain up to 0.200 mg/l of mercury (9). Caustic, one product of the chloralkali process, has been reported to contain very high levels of mercury (Tables 53 and 54).

Table 53. Mercury Content of Caustic (NaOH) from Chloralkali Plants.

Mercury (mg/l)	Reference
0.2–13.6	9
125–150	10
7.0	11
10.0–42.0	11

Table 54. Mercury in Caustic from Swedish Chloralkali Plants (9).

	Mercury (mg/l)	
Plant	Range	Average
A	1.8–3.3	2.2
B	2.1–3.2	2.6
C	0.6–1.1	0.8
D	0.07–0.33	0.2
E	4.7–9.8	6.4
F	0.9–2.1	1.5
G	4.1–18.8	12.2
H	0.15–0.34	0.24
Overall Range/Average	0.07–18.8	3.26

TREATMENT TECHNOLOGY

Many types of mercury treatment technology have been described. The effectiveness, and economics, of treatment by each type of technology depends upon the chemical nature and initial concentration of mercury, the presence of other constituents in the wastewater that may interfere with that specific treatment, and the degree of mercury removal that must be achieved. Among the more common types of treatment are precipitation, ion exchange, adsorption and coagulation, reduction of ionic mercury to elemental form and removal by filtration.

Precipitation

Sulfide addition to precipitate highly insoluble mercury sulfide is the most common precipitation treatment reported. Precipitation may possibly be combined with flocculation and separation by gravity settling, filtration or dissolved air flotation (9). These latter steps improve the removal of the precipitated mercury sulfide, but do not enhance the efficiency of precipitation of the soluble mercury itself. Table 55 presents a summary of available treatment data for sulfide precipitation. For high initial mercury levels, sulfide precipitation will achieve 99.9+% removal at alkaline pH, but even with filtration or activated carbon-polishing treatment, minimum efflu-

Table 55. Precipitation Treatment for Mercury.

Treatment Chemical	Mercury Concentration ($\mu g/l$) Initial	Final	Treatment pH	Additional Treatment	Reference
Sodium Sulfide	na	<3	na	vacuum filter	13
Sodium Sulfide	300–6000	10–125	na	pressure filter	13
Sodium Hydrosulfide	131,500	20	3.0	"filter"	14
Magnesium Sulfide	5,000–10,000	10–50	10–11	—	15
	1,000–50,000	10	na	flocculation + activated carbon	16
Sodium Sulfide					
Sulfide	na	<1000	>9	—	17
Sulfide	na	100–300	na	—	18
Sulfide	na	100	na	1. none	19
		10–20		2. activated carbon	
Sulfide	300–6,000	10–125	5.1–8.2	filtration	20

ent mercury achievable appears to be 10–20 μg/l. Most effective precipitation, with regard to minimizing sulfide dosage, appears to occur in the near-neutral pH range (20). Optimum pH of 8.5 has been suggested (9). In addition to lack of capability to reduce mercury below 10 μg/l, other drawbacks of this method include: 1) the formation of soluble mercury sulfide complexes at high levels of excess sulfide, 2) the difficulty of monitoring excess sulfide levels, and 3) the problem of toxic sulfide residual in the treated effluent (20).

Ion Exchange

Most ion exchange treatment of inorganic mercury involves formation of a negatively charged mercuric chloride complex by addition of chlorine (to oxidize metallic mercury) or chloride salts, and removal of the mercuric chloride complex on an anion exchange resin. Most experience with ion exchange treatment has been with chloralkali plant wastes, which contain high background chloride levels. There is some limited data (20,21) to indicate that best ion exchange treatment results from an initial, roughing stage followed by a second polishing ion exchange unit (Table 56). In wastes where chloride is not high, cation exchange resins are effective. Although information on the cation exchange behavior of mercury is scarce, Amberlite IR-120 and Dowex-50W-X8 are both reported to be effective (29). Resins that contain mercapto groups (R-S-H) such as polythiolstyrene are highly specific for mercuric ion (29).

Regardless of whether anionic resins are used for mercuric chloride complex removal, or cationic resins for mercuric ion removal, best ion exchange treatment for inorganic mercury appears to yield an effluent of 1–5 μg/l. Most effective treatment results from two-stage treatment at neutral to slightly acidic pH.

Most available data relates to treatment of inorganic mercury, and there is relatively little information on organic mercurials. However, efficient treatment of methyl mercury acetate has been reported with Dowex A-1 Chelate (30) and Ionac SRXL resins (31). Ionac SRXL resin is no longer commercially available although a substitute, Srafion NMRR, is reported to have identical properties. (31).

Costs of the Osaka ion exchange process (Table 56) have been reported as $400,000–$575,000 for plants of 300–600 tons chlorine/day capacity, including licensing fee. Applying these costs to plants of 0.018–0.032 MGD, capital costs would be $18,000–$22,000/1000

Table 56. Ion Exchange Treatment for Mercury.

Resin Type	Mercury Form	Treatment pH	Mercury (μg/l)		Additional Treatment	Reference
			Initial	Final		
Mtylon-T	Inorganic	5–6	5,000–25,000	1	—	22
Lewatit	Inorganic	na	na	15	—	23
Anion[a]	Inorganic	7	850	2.5	—	24
Macroreticular	Inorganic	na	10,000	<10	—	25
Anion[a]	Inorganic	na	470	30	—	26
Osaka IE[a]	Inorganic	"acidic"	3,000–10,000	100–150	Prefilter	21
Osaka MR[a]	Inorganic	na	100–150	2–5	Osaka IE Resin	21
Imac-TMR	Inorganic	3	na	5	Carbon Prefilter	27
Anion[a]	Inorganic	5–7	na	100	—	18
Activex	Inorganic	na	10	<5	—	28
Ajinomoto	Inorganic	1.5	60	5	—	20
		6.0	87	3	—	
		11.0	1,800	990	—	
Billingsfors-	Inorganic	6.5	35	1	—	20
Langed Q13		na	na	10–20	absorption filter	19

[a] Mercury removed as mercuric chloride complex, $HgCl_x^{(x-2)}$

gpd. Operating costs are given as $1/1000 gal. Credit for recovered mercury may not be included in this cost analysis, however (32).

A Swedish Company (Aktiebolaget Billingsfors-Langed) has developed an ion exchange system which is similar in concept to the Osaka process (33,34). The strong-base first-stage exchanger is regenerable, while the second stage polisher is not. Removal is as the mercuric chloride complex anion, thus requiring pretreatment of mercurous and organic mercury. Capital costs are reported as $2500/1000 gpd for a 0.040-MGD unit. Treatment levels are equivalent to the Osaka process and the unit has been on the market since 1969.

Coagulation Treatment

Removal of mercury by coagulation has been reported for a variety of mercury-containing wastewaters. Coagulants employed include aluminum sulfate (alum), iron salts and lime. The process has been applied successfully to both organic and inorganic mercury. In pilot plant tests of alum and iron treatment of inorganic and methyl mercury, inorganic mercury was removed effectively only by iron, while neither coagulant provided effective methyl mercury removal (35). Similar results have been reported in other studies, in which it was also found that increasing coagulant dosage to 100–150 mg/l did not improve mercury removal (36). Treatment data for alum and iron coagulation are summarized in Table 57. Effluent levels of mercury achieved by alum treatment range from 1.5–102 μg/l, and by iron treatment from 0.5–12.8 μg/l. Iron appears

Table 57. Alum and Iron Coagulation Treatment for Mercury.

Coagulant Salt	Dosage (mg/l)	Treatment pH	Mercury (μg/l) Initial	Final	Additional Treatment	Ref.
Alum	1000	3	11,300	102	Filtration	37
Alum	100	na	90	11	—	38
Alum	100	na	na[a]	10	—	38
Alum	21–41	6.7–7.2	5.9–8.0	5.3–7.4	Filtration	28
Alum	na	7.0	50	26.5	Filtration	35
Alum	20–30	na	3–8	1.5–6.4	—	36
Alum	20–30	na	3–16[a]	2.3–21.3	—	36
Iron	34–72	6.9–7.4	4.0–5.0	1.5–2.5	Filtration	28
Iron	na	8.0	50	3.5	Filtration	35
Iron	20–30	na	1–17	0.5–6.8	—	36
Iron	20–30	na	2–17[a]	1.2–12.8	—	36

[a] Organic mercury.

to provide better treatment than alum, with a lower effluent limit for mercury of 0.5–3.0 μg/l.

Activated Carbon

The effectiveness of carbon treatment is dependent upon several factors. These include initial form and concentration of mercury, dosage and type of activated carbon, and contact period between the carbon and mercury-containing wastewater. Increasing carbon dosages and increasing contact times improve removal of both inorganic and organic mercury (36). Organic mercury is more effectively removed than inorganic mercury. A review of available treatment data (Table 58) indicates that highest percentage removals (85–99%) result from carbon treatment of more concentrated mercury solutions. However, lowest effluent mercury results from treatment of lower initial mercury concentrations, although the relative efficiency is less. Thus carbon treatment of initial mercury below 1 μg/l yields an efficiency of removal of less than 70%, but effluent mercury below 0.25 μg/l. Carbon treatment of initial mercury of 5–10 μg/l yields about 80% removal and effluent levels of below 2 μg/l. Carbon treatment of initial mercury between 10 and 100 μg/l yields 90 or greater percent efficiency, but highest effluent mercury levels of 20 μg/l or less.

Table 58. Activated Carbon Treatment for Mercury.

Mercury (μg/l)		Additional	
Initial	Final	Treatment	Reference
na	<100	Presettling	39
12–46	<0.5	—	
5.8	0.9	Iron Coagulation	35
9.3	1.6	Alum Coagulation	28
na	0.8	—	36
na	0.2	—	36
100	10–20	Precipitation	19
60–87	4.5–6.0	—	20
0.40–0.81	0.13–0.19	Polymer Added	40
0.5	0.21	Polymer Added	40

Reduction Processes

Inorganic ionic mercury can be converted to the metallic form by reduction, and separated by filtration or other solids separation techniques. A variety of reducing agents are available, including aluminum, zinc, hydrazine, stannous chloride and sodium borohydride. A summary of available treatment data on these processes is presented in Table 59. Although there is much discussion of reduction processes in the literature, there is little actual treatment data presented. The main advantage claimed for reduction is that mercury can be recovered, in the metallic state. However, the data of Table 59 indicate that reduction processes cannot effectively achieve mercury levels below 100 μg/l, and that their use would require second stage polishing by one of the more rigorous treatment methods discussed in this chapter.

A full-scale commercial process based on reduction of mercury by sodium borohydride is marketed as the Ventron process (33,34). The process claims to reduce effluent mercury to 10 μg/l. A 15,000 gal/day plant is in operation at a Wood Ridge, New Jersey mercury chemicals plant. Capital costs are reported as $667/1000 gpd, with operating costs of 6.7–26.7¢/1000 gal treated (34). The unit recovers 1–4 lb/Hg per day, at a treatment cost of $1.00/lb of mercury removed.

Table 59. Performance of Reduction Processes for Mercury Treatment.

	Mercury (μg/l)		Additional	
Reductant Used	Initial	Final	Treatment	Reference
Zinc	5,000–10,000	5–10	—	41
Zinc	1,800	140	(pH 11.5)	20
	12,500	830	(pH 10.0)	
	12,500	750	(pH 6.0)	
	12,500	470	(pH 2.5)	
Zinc	na	600	—	19
Stannous Chloride	2,800	500	—	20
Sodium Borohydride	10,000	220	—	20
	4,000	420		
	26,000	820		
Sodium Borohydride	4,700	200		42
Sodium Borohydride	na	<10	(pH 9–11)	18

Other Processes

A variety of other treatment processes are mentioned in the technical literature, although treatment efficiency data are limited. These processes include solvent extraction with high molecular weight amines (13,43,44), use of silicon alloys for reduction of elemental mercury (45), removal by insoluble starch xanthate-cationic polymer complexation (46,47), adsorption onto ground rubber (48) and wool (49) at sulfhydryl sites, and various proprietary processes (19). Although several of the processes appear promising (Table

Table 60. Other Processes for Mercury Treatment.

Process	Mercury ($\mu g/l$)		Comments	Reference
	Initial	Final		
Solvent Extraction with High Molecular Weight Amines	na	na	99.9% removal	43,44
Silicon Alloys	9,900	0.8–2.6	Mg Fe Si Alloy	45
	9,900	3.6–75.6	Ca Si_2 Alloy	
Starch Xanthate—Cationic Polymer Complex	100,000	3.8	pH 7.0	46,47
Ground Vulcanized Rubber	1,800–11,000	10	170 bed vol. throughput	48
Polyethylenimine Modified Wool	1,000	20	—	49
Cationic Polymer Flocculation	na	2.5–15	—	19
Proprietary—Terraneers, Ltd.	na	10	—	19

60), they are all in the experimental stages and have not been proved in pilot or full-scale application.

SUMMARY

There are many types of treatment technology available for mercury control. The more effective are summarized in Table 61, with lower limits of treatment capability indicated. Among the processes shown, ion exchange, iron or alum coagulation and activated carbon adsorption have the capability to reduce mercury below 10 μg/l. Conventional precipitation systems employing sulfide and flocculating aids seem capable of yielding mercury levels of 10–20 μg/l, based on the information available. Other methods, particularly reduction for small-scale use, are capable of producing very low effluent concentrations of mercury. Many of these methods have not been tested in full-scale application, however.

Table 61. Summary of Treatment Technology for Mercury.

Technology	Lower Limit of Treatment Capability Hg (μg/l)
Sulfide Precipitation	10–20
Ion Exchange	1–5
Alum Coagulation	1–10
Iron Coagulation	0.5–5
Activated Carbon	
High Initial Hg	20
Moderate Initial Hg	2.0
Low Initial Hg	0.25

REFERENCES

1. "Mercury Pollution: Interest in it Now on the Upswing," *Water Wastes Eng.* **8,** A13-A16 (January, 1971).
2. National Industrial Pollution Control Council. *Mercury,* Staff Report, U.S. Government Printing Office 0-409-779 (October 1970).
3. Zugger, P. D. and M. W. Ghosh. "The Effect of Mercury on the Activated Sludge Process," Presented at the 27th Ind. Waste Conf., Purdue University, 1972.
4. Cheremisinoff, P. N. and Y. H. Habib. "Cadmium, Chromium, Lead, Mercury: A Plenary Account for Water Pollution. Part 1. Occurrence, Toxicity and Detection," *Water Sewage Works* **119,** 73-85 (1972).
5. Gurney, W. G. "Mercury Pollution: Michigan's Action Program," *J. Water Poll. Cont. Fed.* **43,** 1427-1438 (1971).

6. D'Itri, F. M. "Mercury in the Aquatic Ecosystem," Technical Report No. 23, (East Lansing, Michigan: Institute of Water Research, Michigan State University, 1972).
7. Irukayama, K. "The Pollution of Minamata Bay and Minamata Disease," *Proc. 3rd Intl. Conf. Water Poll. Res., Munich,* 3:153-180, 1966.
8. Cranston, R. E. and D. E. Buckley. "Mercury Pathways in a River and Estuary," *Environ. Sci. Technol.* **6**, 274-278 (1972).
9. Bouveng, H. O. "Control of Mercury in Effluents from Chlorine Plants," *J. Pure Appl. Chem.* **29**, 3:75-91.
10. Newell, I. L. "Mercury and Other Heavy Metals in Water Supplies," *J. New Engl. Water Works Assoc.* **85** (3), 289.
11. Bouveng, H. O. "The Chlorine Industry and the Mercury Problem," *Modern Kemi (Sweden)* **3**, 45 (1968).
12. Fedonina, V. F., *et al.* "Removal of Mercury from Industrial Waste Water," *Khim. Prom. (USSR)* **48** (1), 72-73 (1972); *Chem. Abstr.* **76**, 267 (1972).
13. Immartino, N. R. "Mercury Cleanup Routes-II," *Chem. Eng.* **82** (3), 36-37 (1975).
14. Sato, Y., *et al.* "Recovery Method for Mercury," *Chem. Abstr.* **77**, 295 (1972).
15. Ishida, T. and K. Aiba. "Recovery Method for Mercury," *Chem. Abstr.* **77**, 295 (1972).
16. Fujii, M. "Treatment of Mercury in Waste Water," *Chem. Abstr.* **78**, 256 (1973).
17. Dean, W. E., *et al.* "Mercury Removal from a Brine Solution," *Chem. Abstr.* **77**, 242 (1972).
18. Cheremisinoff, P. N. and Y. H. Habib. "Cadmium, Chromium, Lead, Mercury: A Plenary Account for Water Pollution. Part 2—Removal Techniques," *Water Sewage Works* **119**, 46-51 (August, 1972).
19. "Industrial Waste Study—Mercury Using Industries," U.S. EPA Report 805/25—18,000 HIP 07/71 (1971).
20. Perry, R. "Mercury Recovery from Contaminated Waste Water and Sludges," U.S. EPA Report 660/2-74-086 (December, 1974).
21. Gardiner, W. C. and F. Munoz. "Mercury Removed from Waste Effluent via Ion Exchange," *Chem. Eng.* **78** (19), 57-59 (1971).
22. Fedonina, V. F., *et al.* "Removal of Mercury from Industrial Waste Waters," *Chem. Abstr.* **26**, 267 (1972).
23. Oehme, C. "Purification of Mercury-Containing Waters and Waste Waters by Lewatit Ion Exchangers," *Vom Wasser (German)* **38**, 345-56 (1971).
24. Yokota, N. "Removal of Mercury from Waste Water," *Chem. Abstr.* **76**, 222 (1972).
25. Egawa, H. "Selective Absorption Resins. 3. Recovery and Removal of Mercuric Ion with Macroreticular Chelate Resins Containing

Polyethylenepolyamino Groups," *Chem Abstr.* **75,** 252-253 (1971).
26. Oehme, C., *et al.* "Removal of Mercury from Industrial Wastes," *Chem. Abstr.* **77,** 271 (1972).
27. Rosenzweig, M. D. "Mercury Cleanup Routes-1," *Chem. Eng.* **82** (2), 60-61 (1975).
28. Logsdon, G. S., *et al.* "Removal of Heavy Metals by Conventional Treatment," In *Proc. 16th Water Qual. Conf.*, University of Illinois, (February 12-13, 1974).
29. Roesmer, J. "Radiochemistry of Mercury," NAS-NS-3026 (Rev.), National Academy of Sciences—National Research Council (September, 1970).
30. Muzzarelli, R. and A. Isolati. "Methyl Mercury Acetate Removal from Water by Chromatography on Chelating Polymers," *Water Air Soil Poll.* **1,** 65-71 (1971).
31. Law, S. L. "Methyl Mercury and Inorganic Mercury Collection by a Selective Chelating Resin," *Science* **174,** 285-287 (1971).
32. Gardiner, W. G. and F. Munoz. "Mercury Removed from Waste Effluent Via Ion Exchange," *Chem. Eng.* **78,** 57-59 (August, 1971).
33. Cheremisinoff, P. N. and Y. H. Habib. "Cadmium, Chromium, Lead, Mercury: A Plenary Account for Water Pollution. Part 2. Removal Techniques," *Water Sewage Works* **119,** 46-51 (1972).
34. Rosenzweig, M. D. "Parring Mercury Pollution," *Chem. Eng.* **78,** 70-71 (February, 1971).
35. O'Connor, J. T. "Removal of Trace Inorganic Constituents by Conventional Water Treatment Processes," *Proc. 16th Water Qual. Conf.* University of Illinois (February 12-13, 1974).
36. Logsdon, G. S. and J. M. Symons. "Mercury Removal by Conventional Water Treatment Techniques," *J. Amer. Water Works Assoc.* **651,** 554-562 (1973).
37. Honda, A., *et al.* "Mercury Treatment in the Wastes and Waste Water from the Fluorescent Lamp Industry," *Chem. Abstr.* **78,** 220 (1973).
38. Hellestan, C. J. S., *et al.* "Compositions and Method for Precipitating or Binding Various Forms of Mercury Occurring in Contaminated Water, Such as Lakes," *Chem. Abstr.* **77,** 296 (1972).
39. Suzuki, G., *et al.* "Mercury Pollution Control at Mercury-Process Sodium Chloride Electrolysis Plants. 2. Removal of Mercury from the Solution," *Chem. Abstr.* **78,** 216 (1973).
40. San. Sci. Div., Military Tech. Dept. Progress Report, "420 GPH Laundry Wastewater Treatment Unit," (Ft. Belvoir, Virginia: U.S. Army Mobility Equipment R&D Center, January 1974).
41. Waltrich, P. F. "Apparatus for the Removal of Mercury from Effluent Streams," *Chem. Abstr.* **78,** 235 (1973).
42. Coulter, M. O. "Removal of Mercury from Spent Brine," *Chem. Abstr.* **78,** 226 (1973).

43. Moore, F. L. "Removal of Mercury and Other Toxic Metals from Plant Effluent Solutions by Solvent Extraction." in *Ecology and Analysis of Trace Contaminants,"* Progress Report ORNL-NSF-EATC-1, (Oak Ridge, Tenn.: Oak Ridge National Laboratory, March 1973).
44. Moore, F. L. "Solvent Extraction of Mercury from Brine Solution with High-Molecular-Weight Amines." *Environ. Sci. Technol.* **6** (6), 525-529 (1972).
45. McKaveney, J. P., *et. al.* "Removal of Heavy Metals from Water and Brine Using Silicon Alloys," *Environ. Sci. Technol.* **6** (13), 1109-1113 (1972).
46. Swanson, C. L., *et al.* "Mercury Removal from Wastewater with Starch—Xanthate Cationic Polymer Complex," *Environ. Sci. Technol.* **7** (7), 614-618 (1973).
47. Winz, R. E., *et al.* "Heavy Metal Removal with Starch Xanthate-Cationic Polymer Complex," *J. Water Poll. Cont. Fed.* **46** (8), 2043-2047 (1974).
49. Freeland, G. N., *et al.* "Adsorption of Mercury from Aqueous Solutions by Polyethylenimine-Modified Wool Fibers," *Environ. Sci. Technol.* **8** (10), 943-946 (1974).

15

TREATMENT TECHNOLOGY FOR NICKEL

MAJOR INDUSTRIAL SOURCES

Wastewaters containing nickel originate primarily from metal industries, particularly plating operations. High levels of nickel have also been reported in wastes from silver refineries (1). In addition, basic steel works and foundries, motor vehicles and aircraft industries, and printing are potential sources of nickel (2). Nickel concentrations reported for a variety of wastewater sources are summarized in Table 62. Wastewaters resulting from pickling and plating of nickel and nickel alloys pose far more problems in treatment than is the case for other plating metals such as copper. These difficulties result from the wide variety of nickel alloys, and the variety of pickling liquors used. Acids used in pickling can include sulfuric, nitric hydrochloric and hydrofluoric (16). Alkaline liquors are also used for pickling some alloys. Acid and sulfamate nickel plating baths generally contain nickel at 40,000–90,000 mg/kg of solution, while electrodeless process baths are of weaker concentration, containing 7000–7500 mg/kg of solution (6). Generally, plating rinse waters containing about 1% of the plating bath concentration

Table 62. Summary of Nickel Concentrations Reported in Wastewaters.

Source	Nickel Concentration (mg/l)		Reference
	Range	Average	
Tableware Plating			
Silver bearing waste	0–30	5	3
Acid waste	10–130	33	3
Alkaline waste	0.4–3.2	1.9	3
Metal Finishing			
Mixed wastes	17–51	—	4
Acid wastes	12–48	—	4
Alkaline wastes	2–21	—	4
Small parts fabrication	179–184	181	5
Combined degreasing, pickling and Ni dipping of sheet steel	3–5	—	6
Brass pickling	—	3	7
Business Machine Manufacture			
Plating wastes	5–35	11	3
Pickling wastes	6–32	17	3
Plating Plants			
4 different plants	2–205	—	8
Rinse waters	2–900	—	9
Large plants	—	25	10
5 different plants	5–58	24	11
Large plating plant	88 (single waste stream)	—	12
	46 (combined flow)	—	
Automatic plating of zinc base castings	45–55	—	6
Automatic plating of plastics	30–40	—	6
Manual barrel and rack	15–25	—	6
Other			
Mine drainage	0.19–0.51	—	13
Acid mine drainage	0.46–3.4	—	14
Gold ore extraction	—	6.5	15

(11). Industries concerned primarily with copper or brass plating or processing may also contain nickel in plant wastewaters, but generally at low levels, 5 mg/l or less. In addition, nickel concentrations ranging from 0.46 to 3.4 mg/l have been reported in acid mine drainage.

Nickel exists in waste streams predominantly as the soluble ion. In the presence of complexing agents such as cyanide, nickel may exist in an extremely stable soluble complexed form. The formation of nickel-cyanide complexes interferes with both cyanide and nickel treatment, and the presence of these species may be responsible for high levels of both cyanide and nickel in treated wastewater effluents.

TREATMENT TECHNOLOGY

Regardless of the type of treatment employed, the level of treatment required and associated costs can often be reduced by good housekeeping practices to reduce accidental loss, spillage, or leaks of plating or pickling solutions containing nickel. Process modifications to reduce the major contributor, dragout loss of plating bath solution into rinse waters, have been successful. As one example, a nickel plating system of 1250–gal capacity, which initially used 270 gph of rinse water, by process changes reduced this value to 60 gph (12).

Usually, total flow reduction is an integral part of recovery systems, to reduce capital costs. Countercurrent flow multiple-rinse tanks have been used to allow direct recovery of nickel plating solution dragout. The first (and therefore most concentrated) rinse solution may serve as plating bath solution makeup, to compensate for evaporation from the hot plating bath. Use of four sequential rinse tanks resulted in a nickel concentration of only 0.9 mg/l in the final rinse (17).

Waste treatment facilities can be expected to cost 10% or less of the installed cost of the associated industrial process equipment. Depending upon the industrial process, wastewater treatment method of choice, and size of operation, capital costs totaling 3–5% appear to be most common.

Treatment processes can be classed as destructive, based upon nickel precipitation and disposal, and recovery, which can be accomplished by several means including selective precipitation. The destructive process may be preceded by a separate process which results in concentration of the waste prior to precipitation. Recovered nickel is normally recycled through the metal processing operations, but occasionally may be marketed to reclaimers.

Precipitation

The formation and precipitation of nickel hydroxide is generally the basis for destructive treatment of nickel-bearing wastes. Precipitation with carbonate is most commonly associated with recovery systems. Nickel forms insoluble nickel hydroxide upon addition of lime. The nickel hydroxide has a minimum theoretical solubility of 0.01 mg/l at pH 10 (18). Precipitation is most effective at high pH, although little efficiency is gained about pH 10. Therefore, close pH control is not essential other than from considerations of the chemical costs associated with pH adjustment. Table 63 sum-

Table 63. Summary of Effluent Nickel Concentrations after Precipitation Treatment.

Source	Nickel Concentration mg/l		Comments	Reference
	Initial	Final		
Tableware Plating	21	0.09-1.9	$FeCl_3$ + Sand filtration	3
Appliance Manufacturing	35	0.4	—	4
Office Machine Manufacturing	39	0.17	Polyelectrolyte, 1-2 mg/l	19
Nonferrous Metal Works	—	0.05-0.13	6-hr settling	20
	—	0.05-0.13	pH = 8.5–9.0	21
	—	15	pH = 8.1	22
Plating	46	0.8	6-hr detention in circular clarifier	12
Record Changer Manufacturing	—	0.1-0.2	—	23

marizes full-scale results of lime precipitation. Effluent nickel values greater than 1 mg/l appear to result from improper pH control (3,22).

Kantawala and Tomlinson (24) have reported the reduction of 100 mg/l of nickel to 1.5 mg/l at pH 9.9, by addition of 250 mg/l lime and resultant formation of nickel hydroxide. Figure 33 shows the relationship between nickel removal and lime dose for their laboratory studies. Maruyama, *et al.* (25), in pilot plant studies (15 gpm) added heavy metals to municipal waste at a concentration of 5 mg/l. Lime addition at 260 mg/l (pH 10.0) plus 20 mg/l ferrous sulfate yielded 0.35 mg/l Ni after sedimentation plus mixed media filtration. Lime addition alone at 600 mg/l (pH 11.5) yielded residual nickel of 0.15 mg/l, after sedimentation and filtration.

Effective treatment may depend to a large extent on the precipitate solids separation efficiency. Continuous treatment with gravity clarifiers has been claimed to yield residual total nickel concentrations below 2 mg/l, with lower values possible by coagulation polymer addition (26). There is much data to demonstrate the importance of effective solids removal and further, demonstrate the better performance of gravity sand filtration over pressure filtration. However, filtration may not be economical, relative to properly designed clarifiers which can achieve equivalent solids separation.

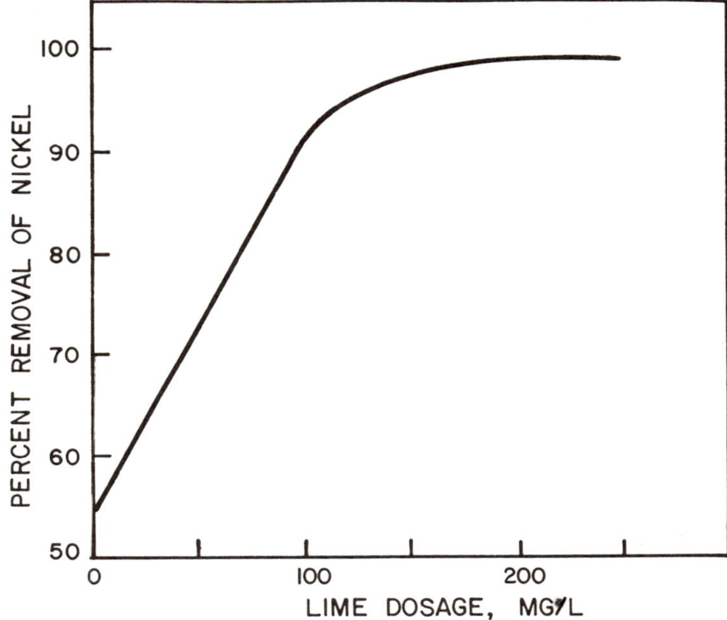

Figure 33. Treatment efficiency for nickel removal by chemical precipitation with lime. [From Kantawala and Tomlinson (24), courtesy of *Water and Sewage Works.*]

The integrated or "Lancy" process (27) is designed to treat plating (or any process bath) dragout solution directly, instead of diluting it into a rinse water and treating the larger volume. The treatment bath solution is continuously circulated from the vat to a reservoir, with sufficient detention time to allow settling of the metal precipitate. Generally, this is the hydroxide for mixed metal wastes. If the waste contains a single metal, other salts may be precipitated, especially if recovery of the metal is desired. For nickel, carbonate baths are used for chemical recovery (28,29). The treatment bath normally contains a solution of soda-ash plus caustic, at pH 8.5–9.5. However, recent studies on carbonate precipitation of nickel and other heavy metals indicate that nickel carbonate precipitate is not formed within the typical detention times of most treatment plants, and that in fact the nickel hydroxide forms preferentially, even in the presence of high carbonate (30). Thus, precipitation by carbonate treatment provides no more effective nickel removal than by hydroxide precipitation.

Sludge removal is provided in the integrated process reservoir, and makeup chemicals are added to the reservoir in proportion to the amount of precipitate formed. Both batch and continuous recovery systems have been developed. If the total recovered precipitate is less than 1 ton/yr, recovery is advised on a weekly batch basis (28). An integrated treatment system, designed into a new plant, treated for iron, zinc and nickel at pH 9.0 and removed sludge once every 2–3 months to sludge filter beds (31). The beds were cleaned 1–2 times/year. This system allowed 85% water recirculation (15% blowdown to control dissolved solids) and the system is claimed to have reduced both plant size and treatment costs.

Subsequent rinses from the integrated chemical bath dragout contain primarily salts derived from the precipitation vat. However, batch neutralization systems are still likely to be required for treatment of contaminated concentrated plating or pickling baths (32). Furthermore, it has been suggested that integrated systems are not economical for plating plants doing odd jobs which produce a combined waste flow of less than 1000 gal/h (33).

The use of the integrated process for nickel treatment for one nickel chloride/nickel sulfate electroplating process resulted in re-

Table 64. Capital and Operating Cost Data for Nickel Waste Precipitation Processes.

Process	Flow (MGD)	Capital Costs ($/1000 gpd)	Operating Cost (¢/1000 gal)	Reference
1. Lime, continuous (included Cr reduction)	0.450	1108 (1966)	80	4
2. Continuous	0.178	953	138	26
Integrated	0.178[a]	270[b]	22[b]	26
3. Continuous	0.480	461	12.3	29
4. Integrated ($NiSO_4$ recovery)	0.480	275	7.0[c]	29
5. Integrated (no recovery)	0.240	173	9.0	29
6. Integrated (CN destruction, Cr reduction, no recovery)	0.480	96	12.7[d]	29
7. Integrated (no recovery)	0.098	223	11.3	29

[a] essentially complete recycle.
[b] based on 0.178 MGD.
[c] does not include credit for nickel recovery value, at 9.0¢/1000 gal treated.
[d] reduced to 7.7¢ by process change to NaOH-Cl_2, and greater recycle.

duction of nickel from an initial range of 5.0–14.2 mg/l to an effluent value of 0.20–0.95 mg/l. The treatment bath contains soda-ash only, at pH 8.5–9.5 (34).

Limited application of sulfide precipitation has been reported for nickel wastewaters (13,15). In one instance, gold ore extraction wastewater, initially at 6.5 mg/l nickel, was reduced to 2.1 mg/l after treatment with sodium sulfide at pH 7 (15). Use of alum at pH 7–8 with the same waste reduced nickel to 0.5 mg/l, however, and of lime or caustic at pH above 10 to below detectable limits.

In a second application of sulfide precipitation, this time to acid mine drainage, nickel levels of 0.19–0.51 mg/l were reduced to 0.13–0.29 mg/l at pH 5.5–6.5. Best treatment occurred at the higher pH (13).

Table 64 summarizes reported cost data for precipitation processes treating wastes that contain nickel ion as one of the contaminants. In addition, cost estimates (*ca* 1958) for generalized neutralization precipitation treatment systems based on a design flow of 0.720 MGD have been reported as follows (12):

	Capital Cost ($/1000 gpd)	Operating Cost (¢/1000 gal)
Batch Treatment, no flow separation	643	19.4
Batch Treatment, flow separation	789	16.6
Continuous Treatment, no flow separation	564	18.6
Continuous Treatment, flow separation	874	15.3

These figures suggest that the lower operating cost advantage of treating separate waste streams can be offset by increased capital costs. For waste flows of greater than 2000 gpm, a continuous treatment process is generally most economical. For flows of less than a few hundred gallons per week, off-site disposal by contract is often best.

Ion Exchange

Recovery of nickel can be accomplished by ion exchange and has been used for some time. Recovery of nickel from process waste streams is more attractive than for some other metals, due to the high value of nickel. A recent report (9) estimates rinse water recovery to be \$0.80–\$7.00/1000 gal for nickel concentrations ranging from 150–900 mg/l.

Cation-anion exchange has been used to treat nickel sulfate plating bath rinse water containing 870 mg/l nickel (35). Complete nickel removal was reported, and the reclaimed water was acceptable for reuse in the rinse tank. The unit processed 1300 gal/hr. Ion exchange treatment requires waste stream segregation, particularly in the presence of cyanide, so that the nickel regenerant can be recovered by precipitation. Nickel-cyanide permanently contaminates ion exchange resins (32).

A second plant employed a scheme of cation exchange for nickel removal, followed by weak base anion exchange (36). Reclaimed concentrations of 60,000 mg/l nickel plus 24,000 mg/l sulfuric acid (the regenerant) were achieved upon regenerating the exchange resins. Chemical costs were given as approximately \$0.52/lb of nickel recovered (36).

With mixed metal wastes where recovery cannot be practiced, it has been demonstrated that for small waste volumes (35,000 gal/day), ion exchange may be more than twice as costly as chemical precipitation, in terms of operating plus capital costs (37). Further, normal regeneration techniques usually produce a dilute acidic effluent for which further recovery or treatment is difficult. In one application of ion exchange to treat a dilute wastewater containing nickel (7–12 mg/l) and zinc (2–6 mg/l), the regeneration effluent at pH 1.5–2.0 contained only 180–250 mg/l nickel, with other metals at equally low concentration (38). In a comparison of two nearly identical plating departments, where one employed ion exchange treatment, and the other continuous precipitation, capital costs were \$1075/1000 gpd vs \$461/1000 gpd for precipitation, and operating costs were 18.4¢/1000 gpd vs 12.3¢/1000 gal for precipitation (29). The ion exchange unit recycled 90% of the process water.

Evaporative Recovery

To be economical, evaporative recovery requires reduction in rinse water volume, thereby increasing nickel concentrations in the rinse waters to be processed by the recovery system. Countercurrent rinse serves this need. A typical evaporative system employing countercurrent rinse, plus nickel and rinse water recycle, is shown in Figure 34.

As early as 1955, the use of evaporative recovery was reported as an economically feasible process (40). Interestingly, only nickel

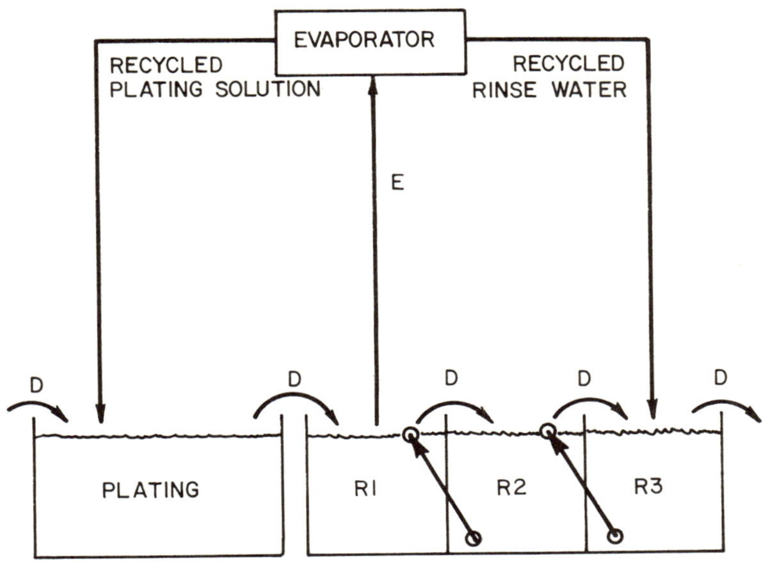

Figure 34. Simple closed loop evaporative recovery system, incorporating countercurrent rinse. [From Culotta and Swanton (39), courtesy of *Plating*.]

was recovered and returned to the plating bath. The evaporated rinse water was not recovered. Capital costs reported were $10,920/1000 gpd for a 1000–gpd system and operating-maintenance costs were $0.85/1000 gal. However, nickel recovery value was estimated at $168/1000 gal treated. Costs for nickel plating solutions have been reported to range from $0.60 to $1.23/gal, and formulas are available for comparison of destructive treatment costs versus evaporative recovery costs (39,41). A commercially available evaporative recovery system, capable of treating 400 gal/hr, has been reported to cost $45,000, and could recover $100,000/yr in chemicals (42). The principal operating cost for evaporative units is steam. This cost makes consideration of double effect or vapor compression evaporators for larger capacity systems essential from a net cost savings viewpoint (43).

Reverse Osmosis

In pilot plant scale studies, reverse osmosis has been found to give 99.6% rejection of plating rinse water nickel ions (44,45). The economics of the process are described in Table 65. The system in-

Table 65. Summary of Cost Estimate for Reverse Osmosis Recovery (45).

	Level A	Level B
RO Unit Capacity (gal/day)	50	500
Installed Capital Cost	$9,000	$37,000
Amortized Capital Cost/1000 gal permeate	$2.68	$1.03
Maintenance Cost/1000 gal permeate	$3.40	$2.05
Total Cost/1000 gal permeate	$6.08	$3.08
Total Cost/day	$21.60	$109.30

cludes three sequential countercurrent rinse tanks. Sufficient makeup is added to the last tank to balance evaporative loss. The contents of the first, most concentrated rinse tank are treated by reverse osmosis, with concentrated nickel solution (12,000 mg/l) returned to the plating tank, and permeate returned to the first rinse tank. Rinse tank nickel concentrations were: first tank 1700 mg/l; second tank 50 mg/l; and third tank 12 mg/l (45).

In a comparison study of reverse osmosis systems application to nickel wastewater treatment, the relative efficiency of recovery (percent nickel rejection) was found to be dependent upon both type of membrane and module used (46). For a nickel sulfamate bath rinsewater, rejection of nickel ranged from 84–98%, while a nickel fluoborate bath rinse yielded rejections of 70–95%.

SUMMARY

The primary sources of nickel wastes are plating and metal processing industries. Thus nickel treatment is frequently combined with treatment for other contaminants, and specific costs of nickel removal are difficult to isolate. Destructive treatment for nickel by lime precipitation is capable of reducing residual nickel concentrations to less than 0.5 mg/l. Costs and efficiency of treatment can be substantially reduced by process control and modification.

Implementation of process modifications also makes nickel recovery an attractive proposition when waste stream separation is practiced, as nickel is a relatively valuable metal. Several recovery systems have been demonstrated to be economically feasible, including precipitate recovery directly or after ion exchange, and evaporative recovery. For hot nickel plating baths, process modifications have led to direct nickel recovery, by return of rinse water to the plating bath.

REFERENCES

1. Banerjee, N. G., and T. Banerjee. "Recovery of Nickel and Zinc from Refinery Waste Liquors: Part I—Recovery of Nickel by Electrodeposition," *J. Sci. Ind. Res.* **11**B, 77-78 (1952).
2. Dean, J. G., F. L. Bosqui and K. H. Lanouette. "Removing Heavy Metals from Waste Water," *Environ. Sci. Technol.* **6**, 518-522 (1972).
3. Nemerow, N. L. *Theories and Practices of Industrial Waste Treatment* (Reading, Massachusetts: Addison-Wesley Publishing Co., Inc., 1963).
4. Anderson J. S. and E. H. Iobst, Jr. "Case History of Wastewater Treatment in a General Electric Appliance Plant," *J. Water Poll. Cont. Fed.* **40**, 1786-1795 (1968).
5. McElhaney, H. W. "Metal-Finishing Wastes Treatment at the Meadville, Pa., Plant of Talon, Inc.," *Sewage Ind. Wastes,* **25**, 475-482 (1953).
6. Lowe, W. "The Origin and Characteristics of Toxic Wastes with Particular Reference to the Metals Industries," *Water Poll. Cont. (London)*, 270-280 (1970).
7. Jester, T. L. and T. H. Taylor. "Industrial Waste Treatment at Scovill Manufacturing Company Waterbury, Connecticut," Presented at 28th Ind. Waste Conf., Purdue University, (May 2, 1973).
8. Wise, W. S. "The Industrial Waste Problem: IV. Brass and Copper Electroplating and Textile Wastes," *Sewage Ind. Wastes* **20**, 96-102 (1948).
9. *State of the Art: Review on Product Recovery,* Water Pollution Control Research Series 17-7 ODJW 11/69, (Washington, D.C.: U.S. Dept. of the Interior, 1969).
10. Pinkerton, H. L. "Waste Disposal," In *Electroplating Engineering Handbook,* 2nd ed., A. Kenneth Graham, Ed.-in-chief, (New York: Van Nostrand Reinhold Co., 1962).
11. Battelle Columbus Laboratories. "An Investigation of Techniques for Removal of Cyanide from Electroplating Wastes," Water Pollution Control Research Series 12010 EIE 11/71 (Washington, D.C.: Environmental Protection Agency, 1971).
12. Chalmers, R. K. "Pretreatment of Toxic Wastes," *Water Poll. Cont. (London)* 281-291 (1970).
13. Larsen, H. P., J. K. P. Shou and L. W. Ross. "Chemical Treatment of Metal-Bearing Mine Drainage," *J. Water Poll. Cont. Fed.* **45** (8), 1682-1695 (1973).
14. Hill, R. D. "Control and Prevention of Mine Drainage," Presented at the Environmental Resources Conference on Cycling and Control of Metals, October 31-November 1-2, 1972, Battelle Columbus Laboratories, Columbus, Ohio.
15. Rosehart, R. and J. Lee. "Effective Methods of Arsenic Removal from Gold Mine Wastes," *Can. Mining J.* 53-57 (June, 1972).
16. Jackson, D. V. "Metal Recovery from Effluents and Sludges," *Metal Finishing J.* **18** (211), 235-242 (1972).
17. Barnes, G. E. "Disposal and Recovery of Electroplating Wastes," *J. Water Poll. Cont. Fed.* **40**, 1459-1470 (1968).

18. Jenkins, S. H., D. G. Knight and R. E. Humphreys. "The Solubility of Heavy Metal Hydroxides in Water, Sewage and Sewage Sludge I. The Solubility of Some Metal Hydroxides," *Internat. J. Air Water Poll.* **8**, 537-556 (1964).
19. Hansen, N. H. and W. Zabban. "Design and Operation Problems of a Continuous Automatic Plating Waste Treatment Plant at Data Processing Division, IBM, Rochester, Minnesota," *Proc. 14th Purdue Ind. Waste Conf.* 227-249 (1959).
20. Stone, E. H. F. "Treatment of Non-ferrous Metal Process Waste at Kynoch Works, Birmingham, England," Proc. 22nd Ind. Waste Conf. 848-865 (1967).
21. Stone, E. H. F. "Treatment of Non-Ferrous Metal Process Waste," *Metal Finishing J.* **18** (212), 280-290 (1972).
22. Tupps, C. C. "Treatment of Wastes for Automobile Bumper Finishing," *Ind. Water Wastes* **6**, 111-114 (1961).
23. "Effluent Treatment at BSR," *Metal Finishing J.* **17**, 248 (1971).
24. Kantawala, D. and H. D. Tomlinson. "Comparative Study of Recovery of Zinc and Nickel by Ion Exchange Media and Chemical Precipitation," *Water Sewage Works* **111**, R281-R286 (1964).
25. Maruyama, T., S. A. Hannah and J. M. Cohen. "Removal of Heavy Metals by Physical and Chemical Treatment Processes," Presented 45th Annual Conf. Water Poll. Cont. Fed. (1972).
26. Lancy, L. E., W. Nohse and D. Wystrach. "Practical and Economic Comparison of the Most Common Metal Finishing Waste Treatment Systems," *Plating* **59**, 126-130 (1972).
27. Lancy, L. E. "An Economic Study of Metal Finishing Waste Treatment," *Plating* **54**, 157-161 (1967).
28. Pinner, R. "Quantitative Recovery of Nickel Dragout Losses by the Integrated Method," *Metal Finishing J.* **14**, 272-273 (1968).
29. Pinner, R. and V. Crowle. "Cost Factors for Effluent Treatment and Recovery of Materials in the Finishing Department," *Electroplating Metal Finishing,* 13-31 (1971).
30. Patterson, J. W., H. E. Allen and J. J. Scala. "Heavy Metals Treatment by Carbonate Precipitation," Presented at 30th Ind. Waste Conf., Purdue University (May 8, 1975).
31. "Trade Plater Moves to New Factory and Goes Automatic," *Prod. Finishing* **36**, 28-32 (1971).
32. Silman, H. "Treatment of Rinse Water from Electrochemical Processes," *Metal Finishing* **69**, 62-66 (1971).
33. Jackson, D. V. "Metal Recovery from Effluents and Sludges," *Metal Finishing J.* **18**, 235-242 (1972).
34. Martin, J. J., Jr. "Chemical Treatment of Plating Waste for Removal of Heavy Metals," U.S. Environmental Protection Agency Report EPA-R2-73-044 (May, 1973).
35. Reents, A. C. and D. M. Stromquist. "Recovery of Chromate and Nickel Ions from Rinse Waters by Ion Exchange," *Proc. 7th Purdue Conf.,* 462-473 (1952).
36. Heidorn, R. F. and H. W. Keller. "Methods for Disposal and Treatment of Plating Room Solutions," *Proc. 13th Purdue Ind. Waste Conf.* 418-426 (1958).

37. Ross, R. D. *Industrial Waste Disposal* (New York: Van Nostrand Reinhold Co., 1968).
38. Phasey, N. W. "Experiences with Ion Exchange Resins for Effluent Treatment," *Prod. Finishing* **25** (11), 28-34 (1972).
39. Cullota, J. W. and W. F. Swanton. "Case Histories of Plating Waste Recovery Systems," *Plating* **57**, 251-255 (1970).
40. Hesler, J. C. "Practical Methods for Treatment of Metal Finishing Wastes," *Plating* **42**, 1019-1029 (1955).
41. Culotta, J. W. and W. F. Swanton. "The Role of Evaporation in the Economics of Waste Treatment for Plating Operations," *Plating* **55**, 957-961 (1968).
42. Besselievre, E. B. *The Treatment of Industrial Wastes*, (New York: McGraw-Hill Book Co., 1969).
43. Culotta, J. M. and W. F. Swanton. "Recovery of Plating Wastes: Selection of Lowest Cost Evaporator," *Plating* **57**, 1221-1223 (1970).
44. Golomb, A. "Application of Reverse Osmosis to Electroplating Waste Water Treatment. Part II. The Potential of Reverse Osmosis in the Treatment of Some Plating Wastes," *Plating* **59**, 316-319 (1972).
45. Golomb, A. "Application of Reverse Osmosis to Electroplating Waste Treatment, Part III. Pilot Plant Study and Economic Evaluation of Nickel Recovery," *Plating* **60** (5), 482-486 (1973).
46. Donnelly, R. G., R. L. Goldsmith, K. J. McNulty and M. Tan. "Reverse Osmosis Treatment of Electroplating Wastes, *Plating* **61** (5), 432-442 (1974).

16

TREATMENT TECHNOLOGY FOR OILY WASTES

Oily waste materials are often measured in terms of their hexane solubility, for purposes of pollution evaluation. Hexane is an organic solvent employed to extract oily compounds from wastewaters. Oily wastes include greases, as well as many types of oils. Grease is not a specific chemical compound, but a rather general group of semi-liquid materials, which may include fatty acids, soaps, fats, waxes and other similar materials extractable into solvent. Unlike some industrial oils which represent precise chemical composition, greases are, in effect, defined by the analytical method employed to separate them from the water phase of the waste (1). The following classification for types of oily wastes has been suggested (2):

1) Light hydrocarbons—includes light fuels such as gasoline, kerosine and jet fuel, and miscellaneous solvents used for industrial processing, degreasing or cleaning purposes. The presence of waste light hydrocarbons may make removal of other, heavier oily wastes more difficult.

2) Heavy hydrocarbons, fuels and tars—includes the crude oils, diesel oils, No. 6 fuel oil, residual oils, slop oil, asphalt and road tar.

3) Lubricants and cutting fluids—oil lubricants generally fall into two classes; nonemulsifiable oils such as lubricating oils and greases, and emulsifiable oils such as "water-soluble" oils, rolling oils, cutting oils and drawing compounds. Emulsifiable oils may contain fat, soap, or various other additives.

4) Fats and fatty oils—these materials originate primarily from processing of foods and natural products. Fats result from processing of animal flesh. Fatty oils for the most part originate from plant products. Quantities of these oils result from processing soybeans, cottonseed, linseed and corn.

INDUSTRIAL SOURCES

There are many industrial sources of oily wastes. Table 66 presents the major categories of industry producing oil and grease-laden waste streams, and lists characteristic types and sources of oily wastes associated with each category. Table 67 summarizes reported oily waste concentrations for many of these industries. By far the three major industrial sources of oily waste are petroleum refineries, metals manufacture and machining, and food processors. Petroleum refineries produce large quantities of oil and oily emulsion wastes. Because of the long history of pollution problems associated with petroleum refining, the American Petroleum Institute (API) has exerted a good deal of effort in developing and publishing methods of oily waste control (*e.g.,* 2,15).

In the metals industry, the two major sources of oily wastes are

Table 66. Industrial Sources of Oily Wastes (2).

Industry	Waste Character
Petroleum	Light and heavy oils resulting from producing, refining, storage, transporting and retailing of petroleum and petroleum products.
Metals	Grinding, lubricating and cutting oils employed in metal-working operations, and rinsed from metal parts in clean-up processes.
Food Processing	Natural fats and oils resulting from animal and plant processing, including slaughtering, cleaning and by-product processing.
Textiles	Oils and grease resulting from scouring of natural fibers (*e.g.,* wool, cotton).
Cooling and Heating	Dilute oil-containing cooling water, oil having leaked from pumps, condensers, heat exchangers, etc.

Table 67. Concentrations of Oily Material in Industrial Wastewaters.

Industrial Source	Oily Waste Conc. (mg/l)	Reference
Petroleum Refinery[a]	40-154	3
Petroleum Refinery[a]	35-178	4
Petroleum Refinery[a]	20	2
Steel Mill		
Hot Rolling	20	2
Cold Rolling	700 (500 free oil)	6
Cold Rolling	60-500	7
Cold Rolling Coolant	2,088-48,742 (2,036-36,664 free oil)	8
Cold Rolling Rinse	113-3,034 (83-2,284 free oil)	8
Food Processing	3830	9
Food Processing (Fish)	520-13,700	10
Metal Finishing	100-5,000	11
Metal Finishing	665	12
Oil Field Brine	25-50	13
Paint Manufacture	1900	9
Aircraft Maintenance	500-1200	9
	(250-500 free oil)	9
Wool Scouring	1,605-12,260	14
Yarn Scouring	2,300-8,160	14

[a] API oil separator effluent.

steel manufacture and metal working. Oily wastes include both emulsified and non-emulsified or floating oils. In steel manufacture, steel ingots are rolled into desired shapes in either hot or cold rolling mills. Oily wastes from hot rolling mills contain primarily lubricating and hydraulic pressure fluids. In cold strip rolling, however, the steel ingot is usually oiled prior to rolling, to lubricate and to reduce rusting. Additional oil-water emulsions are sprayed during rolling to act as coolants. Table 68 presents typical oil usage levels for the cold steel rolling process. After shaping, the steel is rinsed to remove the adhering oil. Rinse and coolant waters from the cold rolling mills may contain several thousand mg/l of oil, of which 25% or more may be emulsified and thus difficult to separate from the wastewater. Emulsified oil in hot rolling mill effluents rarely exceeds 20 mg/l (2). More concentrated oily wastes, such as from batch dumps of spent coolant or lubricating fluid must also be treated.

Metal-working produces shaped metal pieces such as pistons and other machine parts. Oily wastewaters from metal working processes

Table 68. Oil Use in Cold Rolling of Steel.

Manufacturing Process	Oil Dosage, lb/ton steel	Reference
General Use	1.4	6
Coating Oil	1.5–3.0	7
Rolling Oil	3–6	7
Miscellaneous Lubricant	0.3–0.7	7

contain grinding oils, cutting oils, and lubrication fluids. Coolant oil-water emulsions are also employed in many metal-working processes. Soluble and emulsified oil content of wastewaters may vary from 100–5000 mg/l (11).

A third major source of oily wastes, and particularly greases, is the food processing industry. In the processing of meat, fish and poultry, oily and fatty materials are produced primarily during slaughtering, cleaning, and by-product processing. The major grease sources are the rendering areas, in particular from the wet (or steam) rendering process which gives the highest levels, pound per pound of scrap processed, of hexane extractables in food processing waste streams (2). Grease content in meat packinghouse waste streams may run several thousand mg/l (16), and waste from a fish processing plant has been reported to contain 520–13,700 mg/l of fish oil (10).

CURRENT TREATMENT TECHNOLOGY

Treatment of oily waste is similar in concept to treatment of domestic sewage. In domestic sewage treatment a primary level of treatment is employed to separate the easily settleable solids from the liquid—and in treatment of oily waste, primary treatment separates the floatable (free) oils from the water and emulsified oily material. A secondary treatment phase is then required to break the oil-water emulsion and separate the remaining oil and water (17).

Primary Treatment

Primary treatment takes advantage of the difference in specific gravity of oils and greases versus water. The treatment process normally involves retaining the oily waste in a holding tank and allowing gravity separation of the oily material, which is then skimmed from the wastewater surface. Gravity-type separators are the most common devices employed in oily waste treatment. The effectiveness of a gravity separator depends upon proper hydraulic design, and design period of wastewater retention. Longer retention times allow better separation of the floatable oils from the water. The influence of retention time on separator efficiency for a refinery wastewater is indicated in Figure 35. Short detention times

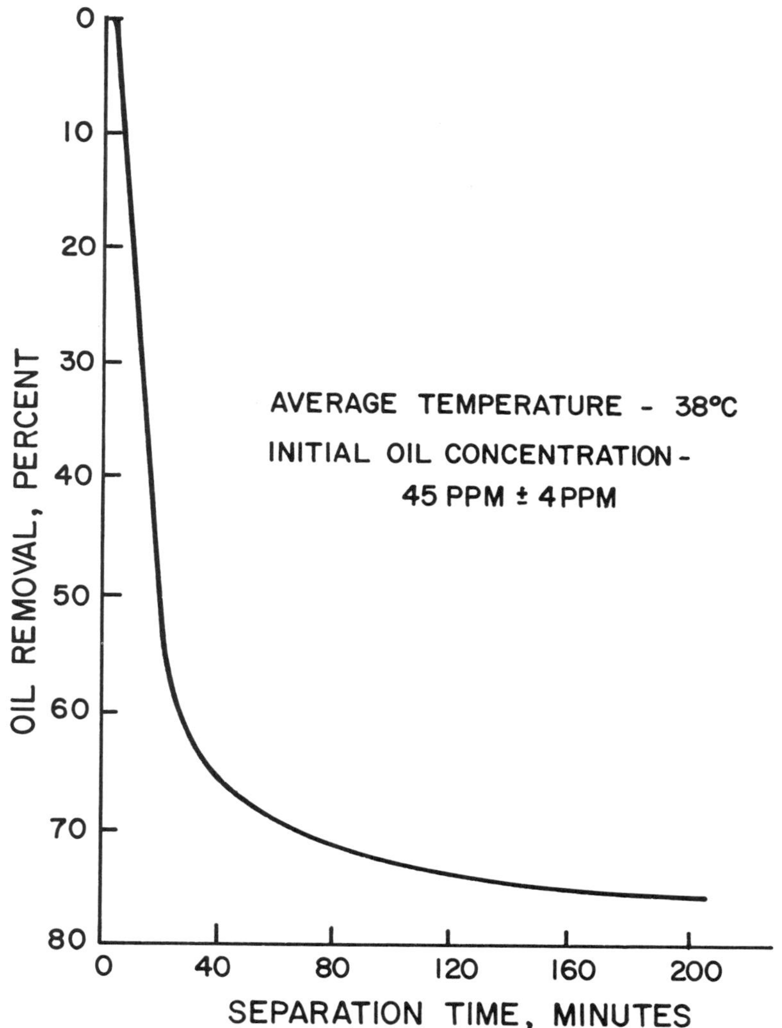

Figure 35. Effect of detention time on oil removal by gravity separation. [From Wallace, *et al.* (18), courtesy of Purdue University Industrial Waste Conference.]

of less than 20 minutes resulted in less than 50% oil-water separation, while more extended holding periods improved oil separation from the waste stream.

Gravity separators are equally effective in removing both greases and nonemulsified oils. The standard unit in refinery waste treatment is the API separator, based upon design standards published by the American Petroleum Institute (15). Separators used for metal and food processing oily wastes operate upon the same principle of floating the oil, and many are designed in a similar fashion to the API process insofar as skimming, retention time, etc. Separators may be operated as batch vats, or as continuous flow-through basins, depending upon the volume of waste to be treated.

Table 69 presents the efficiencies of several oil separation processes, including the API separator. The values given in this table are based upon treatment of petroleum refinery waste. However, efficiencies presented are also generally applicable to oily wastes from other sources undergoing the indicated treatment. For example, 90% separation of free oil has been reported in a holding tank receiving an influent oil concentration of 700 mg/l (470 mg/l free and 230 mg/l emulsified) (6). The waste stream originated from a cold steel rolling mill. The flow stream volume was 3000 gpm. Additional treatment, described subsequently in this chapter, was required for the emulsified oil, to meet an effluent requirement of 15 mg/l of oil. Peoples, *et al.* (20) have reported API separator oil effluent levels, for a refinery waste, ranging from 35–178 mg/l, with an average concentration of 60 mg/l. For a similar refinery waste, API separator effluent oil levels averaging about 20 mg/l have been cited (5).

Table 69. Efficiencies of Oil Separation Processes (19).

Treatment	Source of Influent	Percent Removal: Floating Oil, (%)	Percent Removal: Emulsified Oil (%)
API Separator	Raw Waste	60-99	Not applicable
Air Flotation without Chemicals	API Effluent	70-95	10-40
Air Flotation with Chemicals	API Effluent	75-95	50-90
Chemical Coagulation and Sedimentation	API Effluent	60-95	50-90

Grease content of the waste stream from a meat packing plant was 2850 mg/l, for a flow volume ranging from 1–7 MGD (16). A 90-minute detention period in the gravity separator, and skimming of the floating greases reduced the waste stream grease content by about 75% to approximately 750 mg/l. Grease was recovered for reprocessing and sale. Recovery of skimmed oil or grease from all major types of oily waste is increasingly common, as the value of the recoverable oil is realized. Frequently a substantial savings is possible through recovery or recycle of oily material. If skimmings cannot be reused, they are typically disposed of by burial, lagooning, or incineration. Odor and nuisance-free oil sludge incineration has been reported (21,22).

Table 70. Capital Costs for Refinery Oily Waste Treatment (19).

Process	Flow (MGD)	Capital Cost ($)	Cost ($/MGD)
API Separator	3.0	46,500	15,500
	7.5	105,000	14,000
	15.0	210,000	14,000
Air Flotation	3.0	111,500	37,167
	7.5	222,750	29,700
	15.0	371,250	24,750

Table 71. Annual Maintenance and Operating Costs for Refinery Oily Waste Treatment (19).

Process	Flow (MGD)	M&O Cost ($/Yr)	Cost ($/MGD)
API Separator	3.0	23,000	7,667
	7.5	36,000	4,800
	15.0	55,000	3,667
Air Flotation	3.0	37,000	12,333
	7.5	74,000	12,333
	15.0	117,750	7,850

Capital costs of $1,200,000 have been reported (1966 prices) for an API separator with a capacity of approximately 2000 gpm (17). Refinery oily waste treatment costs have also been reported for both capital and maintenance and operation (19). These data for a typical level of refinery technology are summarized in Tables 70 and 71. Equivalent data, on capital costs for the meat processing industry, and maintenance and operation costs for grease and oil treatment of food industries wastewaters, are summarized in Tables 72 and 73.

Table 72. Capital Costs for Treatment of Meat Processing Oily Wastes (23).

Process	Flow (MGD)	Capital Cost ($)	Cost ($/MGD)
Gravity Separator	0.125	12,000	96,000
	0.54	35,000	64,815
	1.68	250,000	14,881
Air Flotation	0.125	[a]	[a]
	0.54	60,000	111,111
	1.68	150,000	89,285

[a] Not reported.

Table 73. Annual Maintenance and Operating Costs for Treatment of Meat Processing Oily Wastes (23).

Process	Flow (MGD)	M&O Costs ($/Yr)	Cost ($/MGD)
Gravity Separator	0.125	1,000	8,000
	0.54	10,000	18,519
	1.68	18,000	10,714
Air Flotation	0.125	[a]	[a]
	0.54	13,000	24,074
	1.68	30,000	11,905

[a] Not reported.

Secondary Treatment

Unlike primary treatment, which consists only of gravity separation plus skimming, any of several different processes are directed toward breaking the oil-water emulsion which has passed through the primary separator, and separating the demulsified oil from the water phase.

Emulsions may be broken by chemical, electrical, or physical methods. Chemical methods are in widest use for treatment of oily wastewaters. The electrical process is directed toward emulsions containing mainly oil, with small quantities of water. Most oily wastes in contrast are primarily water with lesser amounts of oil. Physical emulsion-breaking methods include heating, centrifugation, and precoat filtration, with the latter two most common. Centrifugation breaks oil emulsions by separating the oil and water phases under the influence of centrifugal force. Centrifugation is best applied to oily sludges, and is generally not used in treatment of the typical dilute oily wastewater stream unless the volume is small (2).

In one treatment process, a mobile heating plus centrifugation process has been applied to treat an oily waste at 10,000 gal/wk (24). The oil is primarily emulsified, resulting from a metal working plant. In the process, magnesium chloride is added, the waste is heated to 95°C for 3–15 minutes, and centrifuged. Oil is reduced from initial levels of 2000–4000 mg/l to effluent values of 22–89 mg/l. The process is claimed to be rapid, and produce an effluent of near-neutral pH (24).

Filtration has been investigated on a 12-gpm pilot scale, for treating a cold rolling steel mill waste containing 230 mg/l of emulsified oil. The mill waste volume was 3000 gpm. The waste stream first passed through an oil-water separator to remove the floating oil (approx. 470 mg/l) by skimming (6). The overall goal of treatment was to produce an effluent containing less than 15 mg/l of oil. Three alternative levels of secondary treatment were evaluated with respect to achieving the required level of oil removal. These were: (a) a high rate sand and gravel filter (HRF), (b) addition of a polyelectrolyte coagulant prior to high rate filtration, and (c) passage of the HRF effluent through a diatomaceous earth filter (DE). Table 74 lists effluent quality (for an average influent oil content of 230 mg/l) versus projected capital and operation costs for the various treatment alternatives. On the basis of the pilot plant results, a full-scale (3000-gpm) treatment plant, which employs treatment alternative (b) above, was constructed in 1970. Other applications of HRF to oily wastes are summarized in Table 75.

Table 74. Filtration Treatment of an Emulsified Oily Waste (6).

Treatment	Effluent Oil (mg/l)	Capital Cost ($)	Operating Costs ($/day)
HRF	20	750,000	300
Coagulant & HRF	10	775,000	372
HRF & DE	5	1,000,000	440

Chemical treatment of an emulsion is usually directed toward destabilizing the dispersed oil droplets, or destroying any emulsifying agents present. The process may consist of rapidly mixing coagulant chemicals with the wastewater, followed by flocculation and flotation or settling. Acidification may also be effective in breaking an oil-water emulsion. Alternative chemical demulsifying processes are (26):

(1) adding of coagulating salts
(2) adding of acids

Table 75. High Rate Filtration Treatment of Oily Wastewaters.

Wastewater Source	Flow (gpm)	Initial Oil (mg/l)		Final Oil (mg/l)		Ref.
		Range	Average	Range	Average	
Petroleum Refinery[a]	12-8	35-178	60	7-17	11	20
Oil Field Brine	77-112	4-203	29.8	3.3-17	7.2	13
Oil Field Brine	4000		9.85		0.775	13
Oil Field Brine		15-30		2-3		25

[a] Results for pilot plant operation.

(3) adding of salts and heating the emulsion
(4) adding of salts and treatment by electricity
(5) adding of acids plus organic cleaving agents (demulgaters).

In a discussion of these processes, it has been emphasized that coagulation with aluminum or iron salts is generally effective for demulsifying oily wastes (27). However, the aluminum or iron may form hydroxide sludges, which are difficult to dewater. Acids generally cleave emulsions more effectively than coagulant salts, but are more expensive and the resultant acid wastewater must be neutralized after oil-water separation. The pH required for demulsification depends upon the nature of the waste, with pH as low as 2–3 reported in one application (28) and a pH of 5.0–5.5 being effective in another (29). Acid pickling wastewater may be employed if available, to break the oily emulsion. The use of waste hydrochloric and sulfuric acids from plating operations has been reported (30). The "salting out" of the emulsifier by adding large quantities of an inorganic salt may create additional pollution problems by significantly increasing the dissolved solids in the effluent. Organic demulgaters are extremely effective demulsifying agents, but due to their high cost are considered impractical at high rinse water flow rates and low oil concentrations (27). Thus acid or coagulant salts find the widest acceptance as demulsifying agents in industry. Frequently, coagulant aids are added to assist in flocculating the coagulant particles of oil sludge.

Following destabilization of the oily emulsion, air flotation is commonly employed to separate the oil and water. Air flotation consists of saturating a portion of the wastewater with dissolved air under high pressure. The pressure is then suddenly released, resulting in the formation of thousands of microscopic air bubbles which attach themselves to oil droplets and float them to the surface. This oil-air bubble mixture forms a froth layer at the surface, which is skimmed away. Chemical flocculating agents such as salts of iron and aluminum with or without organic polyelectrolytes, are particularly helpful in improving the effectiveness of the air flotation process.

Expected efficiencies of air flotation, with and without the use of chemical aids, are presented in Table 69. Coagulant aids can double the effectiveness of air flotation in removing emulsified oils. Capital and operating costs of air flotation, as a function of waste stream volume, are presented in Tables 70–73.

In one application of this process, 62% oil removal was achieved by direct air flotation, and 94% removal upon addition of 25 mg/l aluminum sulfate coagulant (17). For an oil refinery wastewater 70% oil removal resulted by air flotation, while addition of polyelectrolyte and bentonite clay increased oil removal to 95% (17). For another refinery, 79% oil removal was obtained by flotation, but was increased to 87% removal with an aluminum sulfate coagulant dosage of 25 mg/l (17). In a third refinery operation, an air flotation efficiency of 70–80% resulted, with up to 90% oil removal upon addition of aluminum sulfate at concentrations of 30–70 mg/l. Lime, added at 75–100 mg/l provided even better removal to 95% (17). Capital cost for the air flotation unit was reported at $360,000 (1966 price); waste volume treated was approximately 200 gpm. Maintenance and operating costs were 11.5¢/1000 gal of waste.

Air flotation treatment has been applied to several industrial wastes, some with and some without chemical demulsifying aids. Treatment data are summarized in Table 76. Best effluents resulted from pretreatment with demulsifiers (*e.g.*, alum), prior to air flotation.

An industrial pilot process achieved 93% grease reduction of a meat-packing waste by air flotation (16). Grease content of the waste was reduced from 1944 mg/l to 142 mg/l, with 300 mg/l of aluminum sulfate added. The process resulted in decreased volumes and qualities of recoverable grease however. Cost of chemicals was 13.4¢/1000 gal of waste treated, and capital costs were estimated at $30,000 for a 1-MGD unit.

Chemical destabilization of oily waste is sometimes followed by flocculation and sedimentation, rather than air flotation and skimming. Table 69 presents expected treatment efficiencies for chemical coagulation and sedimentation. One company treats oily waste

Table 76. Air Flotation Treatment of Oily Wastewaters.

Industrial Source	Coagulant	Oil Concentration (mg/l)		Percent Removal	Reference
		Influent	Effluent		
Refinery	none	125	35	72	9
	100 mg/l alum	100	10	92	
Refinery	none	154	40	74	3
Oil Tanker Ballast Water	100 mg/l alum (+ 1 mg/l polymer)	133	15	89	9
Paint Manufacture	150 mg/l alum (+ 1 mg/l polymer)	1900	0	100	9
Aircraft Maintenance	30 mg/l alum (+ 10 mg/l activated silica)	250–700	20–50	90+	9
Meat Packing	none	3830	270	93	9
Meat Packing	none	4360	170	96	9
Meat Packing	300 mg/l alum	1944	142	93	16
Yarn Scouring	aluminum chlorohydrate	2300–8160	40–248	89–99	15
Wool Scouring	aluminum chlorohydrate	1650–12260	35–287	82–99+	15

from a ball and roller bearing plant by coagulation with sodium carbonate, lime and a polyelectrolyte flocculating agent to achieve grease and oil reduction from 302 mg/l down to 28 mg/l. This exceeds 90% removal, in a process treating 30,000–60,000 gal/day. Coagulation with alum (35 mg/l) followed by sedimentation reduced API separator effluent oil from initial levels of 50–100 mg/l to a final effluent oil content consistently less than 15 mg/l. Treatment was on a refinery waste flow of 300 gpm (31). A machining plant wastewater is treated by sulfuric acid addition to pH 2–3, to break the oil water emulsion. After 6–12 hours of acid digestion, calcium chloride and caustic soda are added to coagulate the oil. By contrast to the similar oil treatment processes, the coagulated material floats. The process reduces oil from an average of 21,225 to 6.0 mg/l (28). Advantage claimed for the process, in addition to efficiency exceeding 99%, is that the floated precipitate can be burned with waste oil in the powerhouse. Capital investment for the treatment system was $1,800,000, for a wastewater volume of 200,000 gpd (28).

Oil and grease laden waters are, on occasion, treated in lagoons and other biological treatment processes. In one instance, the use of an anaerobic lagoon achieved 88% grease removal (32). Influent grease content during four 4-day sampling periods within one month ranged from 425–1270 mg/l, while effluent grease concentrations of 69–147 mg/l were recorded. The lagoon treated 528,000 gal/day, at a capital cost (excluding land) of $44,000.

Treatment of a refinery waste by a treatment process involving API separator-air flotation-biological (oxidation pond) treatment has been reported (3). Effluent from the air flotation unit contained 40 mg/l of oil. Effluent from the oxidation pond, which had a minimum 20-day detention time, contained 18 mg/l or less of oil. Another refinery built a 5-MGD extended aeration activated sludge plant to treat refinery waste (5). The unit receives API separator effluent containing about 20 mg/l oil. At design loadings, the activated sludge unit effluent oil is 2–10 mg/l. Annual operating costs for the unit total 4.8¢/1000 gal. Included in this figure is a somewhat high maintenance cost of 1.7¢/1000 gal. Annual maintenance, at $28,765, represented 1.4% of installed cost.

SUMMARY

Treatment of floating oils is achieved economically and efficiently by gravity separation and skimming. Oil removals as high as 99%

have been reported (19). Treatment of emulsified oil-water mixtures is more complex and costly, and represents a secondary phase of treatment after the primary gravity separation process. Chemical coagulation is effectively employed, with air flotation or high rate filtration to separate the oil and water phases. Effluent oil levels of 10–15 mg/l may be expected for coagulant-assisted air flotation, and of 10 mg/l or less for high rate filtration. Without the use of coagulants, these levels of treatment are not achievable. Removals equivalent to that achieved by air flotation or filtration have also been reported for coagulation and precipitation, and for biological treatment processes.

REFERENCES

1. American Public Health Association. *Standard Methods for the Examination of Water and Wastewater,* 13th ed., New York (1971).
2. American Petroleum Institute. *Industrial Oily Waste Control,* New York.
3. Wigren, A. A. and F. L. Burton. "Refinery Wastewater Control," *J. Water Poll. Cont. Fed.* **44,** 117-128 (1972).
4. "Humble Oil Treats Wastes at Baytown," *Water Sewage Works* **118,** IW4-5 (1971).
5. Rose, W. L. and G. E. Gorringe. "Activated Sludge Plant Handles Loading Variations," *Oil Gas J.* **70** (40), 62-65 (1972).
6. Symons, C. R. "Treatment of Cold Mill Wastewaters by Ultrahigh-Rate Filtration," *J. Water Poll. Cont. Fed.* **43,** 2280-2286 (1971).
7. Donovan, E. J., Jr. "Treatment of Wastewater for Steel Cold Finishing Mills," Water Waste Eng. F22-F25 (November, 1970).
8. Treatment of Wastewater—Waste Oil Mixtures," U.S. EPA Report 12010 EZV 02/70 (1970).
9. Boyd, J. L. and G. L. Shell. "Dissolved Air Flotation Application to Industrial Wastewater Treatment," Presented at 45th Annual Conf., Water Poll. Cont. Fed. (1972).
10. Chun, M. J., R. H. F. Young and N. C. Burbank, Jr. "A Characterization of Tuna Packing Waste," *Proc. 23rd Purdue Ind. Waste Conf.* **33,** 786-805 (1968).
11. Brink, R. J. "Operating Costs of Waste Treatment at General Motors," *Proc. 19th Purdue Ind. Waste Conf.* **19,** 12-16 (1964).
12. Germain, J. E., C. A. Vath and C. F. Griffin. "Solving Complex Waste Disposal Problems in the Metal Finishing Industry," Presented at Georgia Water Poll. Cont. Assoc. Conf. (September, 1968).
13. Wallace, J. T., Jr. and J. W. Brown. "Deep Bed Filtration of Oilfield Produced Water," Presented at 27th Ind. Waste Conf., Purdue University (1972).
14. Evans, B. R. "Treatment of Scouring Liquor by Electroflotation," *Effluent Water Treat. J.* **14** (2), 85-88 (1974).
15. American Petroleum Institute. *Manual on Disposal of Refinery Wastes,* 7th ed., New York (1963).

16. Garrison, V. M. and R. J. Geppert. "Packinghouse Waste Processing Applied Improvement of Conventional Methods," *Proc. 15th Purdue Ind. Waste Conf.* **15**, 207-217 (1960).
17. Quigley, R. E. and E. L. Hoffman. "Flotation of Oily Wastes," *Proc. 21st Purdue Ind. Waste Conf.* **21**, 527-533 (1966).
18. Wallace, A. T., G. A. Rohlich and J. R. Villemonte. "The Effect of Inlet Conditions of Oil-Water Separators at SOHIO's Toledo Refinery," *Proc. 20th Purdue Ind. Waste Conf.* **20**, 618-625 (1965).
19. *The Cost of Clean Water: Vol. III, Industrial Waste Profiles, No. 5—Petroleum Refining* (Washington, D.C.: U.S. Dept. of the Interior, 1967).
20. Peoples, R. F., P. Krishnan and R. N. Simonsen. "Nonbiological Treatment of Refinery Waste Water," *J. Water Poll. Cont. Fed.* **44**, 2120-2128 (1972).
21. Dreier, D. E. and J. W. Walker. "Grease Incineration," *Proc. 19th Purdue Ind. Waste Conf.* **19**, 161-16 (1964).
22. "Fluid Bed Incineration of Petroleum Refinery Wastes," U.S. EPA Report 12050 EKT 03/171 (1971).
23. *The Cost of Clean Water: Vol. III, Industrial Waste Profiles No. 8—Meat Products* (Washington, D.C.: U.S. Dept. of the Interior, 1967).
24. Chalmers, R. K. "Treatment of Wastes from Metal Finishing and Engineering Industries," In *Process in Water Technology,* Vol. I (Elmsford, New York: Pergamon Press, Inc., 1972).
25. Bleakley, W. B. "Wilmington Attacks Water Problems," *Oil Gas J.* 71-72 (September, 1972).
26. Mentens, A. "Treatment of Wastes Originating from Metal Industries," *Proc. 22nd Purdue Ind. Waste Conf.* **22**, 908-925 (1967).
27. Barker, J. E., V. W. Foltz and R. J. Thompson. "Treatment of Waste Oil-Waste Water Mixtures," Presented at Annual Conf., AIChE, Chicago (November, 1970).
28. Vaughn, S. H. and R. S. McCurdy. "Wastewater Treatment at Ford's Windsor Complex," *Ind. Waste* **19** (3), 34-43 (1973).
29. Stoner, L. B. "Waste Treatment Facilities for Jones and Laughlin Steel Corporation Hennepin Works," *Proc. 26th Ind. Waste Conf.* Purdue University (1971).
30. Werner, H. W. "Treatment of Metal Finishing Wastes by Robertshaw Controls Company," Presented at 27th Ind. Waste Conf., Purdue University (1972).
31. McPhee, W. T. and A. R. Smith. "From Refinery Wastes to Pure Water," Hydrotechnic Corp., New York.
32. Saucier, J. W. "Anaerobic Lagoons Versus Aerated Lagoons in the Treatment of Packinghouse Wastes," *Proc. 24th Purdue Ind. Waste Conf.* **24**, 534-541 (1969).

17

TREATMENT TECHNOLOGY FOR pH CONTROL

INDUSTRIAL SOURCES

Industrial wastes of many types, and from a variety of processes, exhibit extreme pH values which can greatly influence the quality and aquatic life of receiving waters. Table 77 lists representative industries, and gives waste pH characteristics associated with each industry. Wastes may be released on a batch basis as acid or alkaline processing vats are dumped, or released as a continuous waste stream resulting from a specific industrial process (*e.g.*, rinsing). The volume and composition of a waste stream containing acidic or basic compounds can be quite variable. Dickerson and Brooks (2) have described the waste stream from a nitrocellulose explosive production line of Hercules Powder Corporation as varying in volume from 3000–9000 gpm with acidity, measured as free sulfuric acid, ranging from 30–1100 mg/l.

In addition to the free acids and bases (H_2SO_4, NaOH, etc.) typically encountered in industrial waste streams, some industries produce "combined" acidic or basic salts, which form weak acids or bases by hydrolysis upon dilution of the waste by receiving water.

Table 77. pH Characteristics of Industrial Wastes (1).

Industrial Process	Waste pH Characteristics
Food and Drugs	
Pickling	Acidic or alkaline
Soft drinks	High pH
Apparel	
Textile	Highly alkaline
Leather processing	Variable
Laundry	Alkaline
Chemicals	
Acids	Low pH
Phosphate and phosphorus	Low pH
Materials	
Pulp and paper	High or low pH
Photographic products	Alkaline
Steel	Mainly acid, some alkaline
Metal plating	Acids
Oil	Acids
Rubber stores	Variable pH
Naval stores	Acid
Energy	
Coal mining	Acid mine drainage
Coal processing	Low pH

Typical of such industrial effluents are those originating from steel mills and other metal processing industries, where acid or alkaline cleaning solutions are employed. Control of effluent and stream pH requires treatment not only for the free acids and bases in the waste, but for the acidic and/or basic salts present as well. Waste treatment processes employed for the free acids and bases are effective in controlling acid and base salts.

CURRENT TREATMENT TECHNOLOGY

There are many acceptable methods for treating acidic or basic wastes. These have been discussed in detail in the technical literature (1-5). Treatment of both types of waste is based upon chemical neutralization, usually to pH 6–9. Methods include 1) mixing acid and alkaline wastes so that the net effect is a near-neutral pH; 2) passing acid wastewaters through beds of limestone; 3) mixing acid wastes with lime slurries or dolomite lime slurries; 4) adding the proper amounts of concentrated caustic soda (NaOH) or soda ash (Na_2CO_3) to acid wastewaters; 6) bubbling waste boiler-flue gas through alkaline wastes; 7) adding compressed CO_2 to alkaline wastes; and 8) adding strong acid to alkaline wastes.

All these techniques are well-established and have been employed in treating acid or alkaline industrial wastes. The chemical for and method of neutralization is selected on a basis of overall cost, as chemical costs vary widely, and the equipment for utilizing various neutralizing agents will differ with the treatment method employed (1).

Process wastes may be treated either continuously or on a batch basis. Usually the flow rate will determine this except where treatment also involves removal of some other contaminant, as when pH neutralization is associated with the precipitation of some toxic metal. Generally, it is only when flow rates exceed 70 gpm that continuous neutralization is utilized (6).

The choice of an acidic reagent for neutralization of an alkaline wastewater is generally between sulfuric acid and hydrochloric acid (3). Sulfuric acid is usually selected because of its lower cost. Hydrochloric acid often has the advantage of soluble reaction end products, which may, however, cause the waste to exceed effluent dissolved solids standards.

Carbon dioxide, primarily due to its greater cost, has been employed much less frequently to neutralize alkaline wastes than have sulfuric or hydrochloric acids. Treatment by flue gas, containing up to 14% by volume CO_2, is more common. Capital and operating costs of a flue gas neutralization system employed on a 2.5-MGD flow have been reported (7). The system, which involved two-stage pH adjustment of effluent from an ammonia stripping tower, consisted of first-stage adjustment from pH 11.0 to 9.6 by compressed and scrubbed stack gas, settling of calcium carbonate precipitate, and second-stage pH adjustment by stack gas to the range 6.8–7.5. Total operating costs were $4.40 per million gallons.

The selection of a caustic agent to neutralize an acid waste is usually between sodium hydroxide (caustic soda, NaOH), sodium carbonate (soda ash, Na_2CO_3) and various limes (calcium oxide compounds). The important factors in selection of a caustic reagent include purchase price, neutralization rate and capacity, storage and equipment costs, and neutralization end products. The costs of the various caustic reagents are summarized in Table 78, based upon 1968 price quotations. Although sodium hydroxide is far more expensive than the other materials, it is frequently selected due to composition uniformity, ease of storage and feedings, rapid reaction rate, and solubility of end products. Sodium carbonate is not as reactive as sodium hydroxide and can produce foaming problems due to release of carbon dioxide (3).

Table 78. Cost Comparison of Commercial Caustic Agents (3).

Agent	Chemical Formula	Price[a] ($/ton)	Basicity Factor[b]	Relative Treatment Cost[c]
Sodium hydroxide	$NaOH$	50.00	0.687	6.82
Sodium carbonate	Na_2CO_3	30.00	0.507	5.53
High calcium hydrate	$Ca(OH)_2$	12.00	0.710	1.58
Dolomite hydrate	$Ca(OH)_2 \cdot MgO$	12.00	0.912	1.23
High calcium quicklime	CaO	10.00	0.941	1.00
Dolomite quicklime	$CaO \cdot MgO$	10.00	1.110	1.32
High calcium limestone	$CaCO_3$	4.00	0.489	0.77
Dolomite limestone	$CaCO_3 \cdot MgCO_3$	4.00	0.564	0.67

[a] Costs are 1968 basis (5).
[b] A measure of the alkali available in the commercial product for neutralization (grams equivalent of CaO per gram of agent).
[c] Based upon one unit of cost for CaO, and the treatment effectiveness (re: "Basicity Factor") of each agent.

Limestone and dolomite limestone are inexpensive. However, they are relatively insoluble in water and require direct contact with acidic wastes. In spite of this they are widely used, sometimes as a pretreatment before alkalis, which give closer pH control. Limestone or lime slurries have a tendency to form sludges of insoluble sulfates, and reaction rates are slow (5).

Operating and capital cost data for simple neutralization of acid/alkaline industrial wastes, based upon 1968 prices, are presented in Figure 36. Capital costs are in terms of thousands of dollars invested versus waste flow rate in gallons per minute. Treatment costs are relatively low, even for the fully automated treatment processes shown in Curve III, Figure 36 (a), as compared with other common industrial waste treatment processes such as oil removal or heavy metal precipitation.

The use of both sulfuric acid and hydrated lime, $Ca(OH)_2$, has been reported in treating a metal finishing waste (9). Both acid and alkaline wastes were produced in the industrial process, and an effluent pH of approximately 9.0 was routinely achieved. When lime is used, it is usually fed as a hydrated lime slurry, due to ease of addition and more rapid reactivity. Most effective neutralization generally results from use of multistage reaction tanks with feedforward pH control. Most effective treatment plants incorporate two or three stages of neutralization.

It has been pointed out that acid wastes which contain iron salts, as most metal plating and ferrous industry wastes do, may require

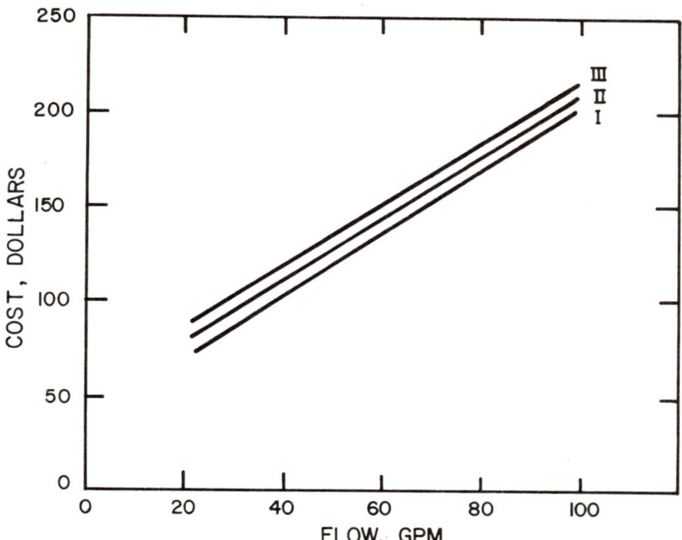

a) Monthly operating plus depreciation costs for pH adjustment (pH 4.0 to 8.3). I-no instrumentation; II-medium instrumentation; III-full instrumentation.

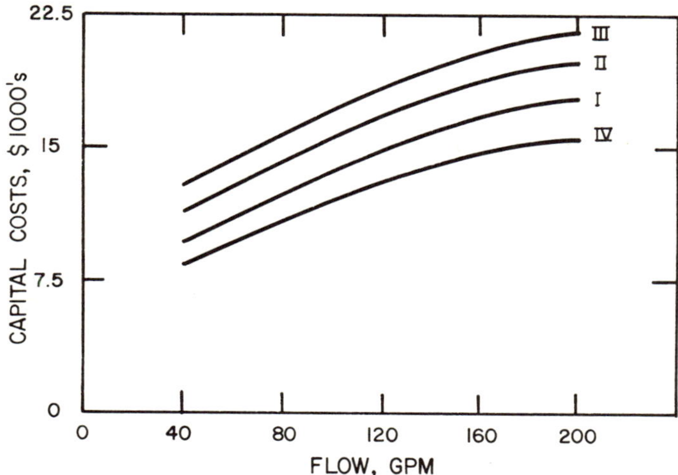

b) Initial capital costs for pH control, including installation. I-manual controls; II-medium instrumentation; III-full instrumentation; IV-package plant.

Figure 36. Costs for single acid/alkaline pH neutralization. [From Zievers, *et al.* (8), courtesy of *Plating.*]

3–10 times the amount of neutralizing caustic agent required by acidic solutions of equivalent pH which do not contain the iron salts (10,11). Waste sulfuric acid, used in pickling (cleaning rust from steel products), typically contains 2–7% (or 20,000–70,000 mg/l) free acid, and 15–22% ferrous sulfate (an acid salt). These salts of weak acids or bases influence the pH of effluents and receiving waters, and exert a chemical demand for additional neutralizing agent.

In discussing the problem of spent pickle liquor from steel processes, Heynike and von Reiche (12) state that the classical method of using the concentrated acid until almost spent, and then neutralizing with lime is outmoded. Dewatering the sludges produced is impractical and large sludge holding lagoons require valuable space. Today, steel producers appear to be turning more to acid recovery and acid regeneration methods, effectively negating the need to treat the concentrated acidic waste (12,13). The necessity for treating dilute acidic rinse water from the pickling operation remains, however. Good rinse water treatment results with limestone slurry, in that it produced a more readily filterable hydroxide sludge than other neutralization chemicals (14,15).

Table 79 summarizes the treatment efficiency and neutralizing agent demand for an acid waste originating from cleaning, plating and other metal finishing operations (4). The acid waste was overneutralized to an alkaline pH, to allow subsequent precipitation of waste heavy metals. Thus the large quantities of lime used resulted from four requirements: 1) Neutralization of the free acids; 2) Neutralization of the combined acids (*i.e.*, acid salts); 3) Overneutralization to an alkaline pH, to achieve conditions suitable for heavy metal-hydroxide precipitation; and 4) The lime required to react with the heavy metals, forming insoluble hydroxides.

Table 79. Acid Waste Treatment (4).

Month (1952)	Waste Vol (gal)	pH		Lime Required (lb)
		Before	After	
July	63,000	1.0	9.9	23,250
August	46,000	1.0	10.1	23,150
September	43,500	1.0	10.0	22,000
October	55,000	1.0	9.9	20,500
November	40,000	1.3	10.1	8,000
December	63,000	1.0	10.2	15,000
Total	310,500			111,900

Because of the high lime demand, operating costs for this acid waste treatment process were extremely high at $26.20/1000 gal. Total treatment costs, including equipment depreciation, was $65.50/1000 gal. Acid treatment costs are high because the acid waste is comparatively concentrated, and considerable lime is required for the neutralization. Capital cost of the acid waste treatment facility was $106,387, exclusive of the waste collection system (4).

Increasing attention is being directed toward neutralization treatment of acid mine drainage, which may have a pH of 2.0 or less (16). A typical neutralization system utilizes sodium carbonate, sodium hydroxide, limestone or lime, with lime being most common. Except for limestone, it is the cheapest reagent (Table 78), and also reacts rapidly. Its disadvantage is that voluminous sludge is produced. Limestone produces a denser sludge, but is not effective above pH 6.5 because of its slow reactivity at higher pH. Two-stage treatment with limestone followed by lime appears to offer the advantages of both reagents (16).

SUMMARY

Neutralization techniques for acid/alkaline wastes are well-established and consist generally of use of sulfuric acid for alkaline solutions, and one of several caustic chemicals for acid waste streams. Costs for treatment of wastes containing only free acids or bases are low, and normally represent only a fraction of the total waste treatment costs of a typical industrial plant. However, if the waste, in addition to free acids, contains other contaminants such as acid or base salts (*e.g.*, ferrous sulfate), or heavy metals that will precipitate simultaneously with pH adjustment, the treatment chemicals demand is sharply increased, as well as the associated cost of waste treatment.

REFERENCES

1. Nemerow, N. L. *Theories and Practices of Industrial Waste Treatment* (Reading, Massachusetts: Addison-Wesley Publishing Co., Inc., 1963).
2. Dickerson, B. W. and R. M. Brooks. "Neutralization of Acid Wastes," *Ind. Eng. Chem.* **42,** 599-605 (1950).
3. Ross, R. D. *Industrial Waste Disposal* (New York: Van Nostrand Reinhold Co., 1968).
4. Cooper, J. E. "How to Dispose of Acid Wastes," *Chem. Ind.* 684-685 (1950).

5. "Water Pollution Control," *Chem. Eng.* **78** (14), 65-75 (1971).
6. Williams, E. T. "pH Neutralization Process Considerations," *Plating* **58,** 967-970 (1971).
7. Evans, D. R. and J. C. Wilson, "Capital and Operating Costs—AWT," *J. Water Poll. Cont. Fed.* **44,** 1-13 (1972).
8. Zievers, J. F., R. W. Crain and F. G. Barclay. "Waste Treatment in Metal Finishing: U.S. and European Practices," *Plating* **55,** 1171-1179 (1968).
9. Hanson, N. H. "Design and Operation Problems of a Continuous Automatic Plating Waste Treatment Plant at the Data Processing Division, IBM, Rochester, Minnesota," *Proc. 14th Purdue Ind. Waste Conf.* **14,** 227-249 (1959).
10. Omya, S. A. "Treatment of Residual Acid Effluent," *Water Waste Treat.* **12,** 27-28 (1968).
11. MacDougall, H. "Waste Disposal at a Steel Plant: Treatment of Sheet and Tin Metal Wastes," *ASCE* Separate No. 493 (September, 1954).
12. Heynike, J. J. C. and F. V. K. von Reiche. "Water and Pollution Control in the Iron and Steel Industry, with Special Reference to the South African Iron and Steel Industrial Corporation," *Water Poll. Cont. (London)* 569-573 (1969).
13. Dempster, J. H. and H. M. Gomaa. "Recycling of Waste Acid from Steel Pickling Operations," *Ind. Waste* **19** (2), 26-27 (1973).
14. Melzer, S. F. and T. L. Taubken. "New Process Treats Acid Rinse Waters," *Water Wastes Eng.* **8** (1), F6-F8 (1971).
15. Barker, J. E., S. F. Melzer and R. D. Snook. "Limestone Treatment of Rinse Waters from Hydrochloric Acid Pickling of Steel," *Proc. 26th Ind. Waste Conf.*, Purdue University (1971).
16. Mill, R. D. "Control and Prevention of Mine Drainage," Presented at Conf. on Control and Recycling of Metals, Battelle-Columbus Laboratories (1972).

18

TREATMENT TECHNOLOGY FOR PHENOLS

INDUSTRIAL SOURCES

The following industries are characteristic sources of phenolic pollutants (1):

Gas works (production)	Explosives
Wood distillation	Coal tar distilling
Oil refineries	Mine flotation wastes
Sheep and cattle dip	Insecticides
Chemical plants	Resin manufacture
Photographic developers	Coke ovens

In addition, aircraft maintenance (2,3), foundry operations (4), Orlon manufacture (5), caustic air scrubbers in paper processing plants (6), rubber reclamation plants (7,8), nitrogen works (9), fiberboard factories, plastic factories, glass production (10), stocking factories (11) and fiber glass manufacturing (12-14) have been reported as contributing phenols to wastewaters. Table 80 summarizes the levels of phenol found in wastes of various industries.

Although described in the technical literature simply as phenols,

Table 80. Levels of Phenol Reported in Industrial Wastewaters.

Industrial Source	Phenol Concentration (mg/l)	Reference
Coke Ovens		
Weak ammonia liquor, without dephenolization	3350-3900	15
	1400-2500	16
	2500-3600	17
	3000-10000	18
	580-2100	19
	700-12000	10
	600-800	20
Weak ammonia liquor, after dephenolization	28-332	21
	10	18
	10-30	16
	4.5-100	9
Wash oil still wastes	30-150	22
Oil Refineries		
Sour water	80-185 (140 ave)	22
General waste stream	40-80	23
Post-stripping	80	24
General (catalytic cracker)	40-50	25
Mineral oil wastewater	100	9
API separator effluent	0.35-6.8 (2.7 ave)	26
General wastewater	30	27
General wastewater	10-70	28
General wastewater	10-100	29
Petrochemical		
General petrochemical	50-600	30
Benzene refineries	210	9
Nitrogen works	250	9
Tar distilling plants	300	9
Aircraft maintenance	200-400	2, 3
Herbicide manufacturing	210	31
(includes chloroderivatives and phenoxy acids)	239-524	32
Other		
Rubber reclamation	3-10	7, 8
Orlon manufacturing	100-150	5
Plastics factory	600-2000	10
Fiberboard factory	150	10
Wood carbonizing	500	10
Phenolic resin production	1600	33
Stocking factory	6000	11
Synthetic phenol, plastics, resins	12-18	34
	369 (after cooling water separation)	
Fiberglass manufacturing	40-400	13

this waste category may include a variety of similar chemical compounds among which are various phenols, chlorophenols and phenoxyacids. In terms of pollution control, reported concentrations of phenol are thus the result of a standard analytical methodology that measures a general group of similar compounds rather than being based upon specific identification of the single compound, phenol (hydroxybenzene).

EXISTING TREATMENT TECHNOLOGY

Treatment technologies for phenolic wastes are available for reduction of all levels of initial phenol concentration, and frequently there is a region of overlap between methods. Both chemical-physical and biological treatment are in successful full-scale industrial use, and high efficiencies of treatment are reported. However, any particular industrial system may encounter difficulties in employing a specific treatment method, depending upon the overall composition of the waste stream. A case in point involves attempting biological treatment on a phenolic waste containing high levels of heavy metals, prior to metal removal. Phenolic wastes often contain large quantities of other waste constituents which require special treatment procedures. This is particularly true for refinery and coke oven wastes. Oil and cyanides are commonly present in large quantities in phenolic wastewaters, and their removal, with attendant economical disadvantages, prior to phenol removal may be required. In general, however, phenol removal is successfully practiced by industry.

Treatment methods for phenol can be best discussed in terms of the phenol concentration in the waste to be treated. It is apparent from Table 80 that industrial wastes may contain an extremely wide range of phenolic material. Therefore, treatment technology is discussed below under three headings—concentrated, intermediate, and dilute phenolic wastewaters. Recycle of process wastewater, with or without phenol removal, is also used with apparent success, and is discussed separately following the treatment technology sections.

Concentrated Wastewaters

High concentrations (above 500 mg/l) of phenol in a waste stream make recovery an attractive economic consideration. Phenol recovery value from coke plant ammonia liquor is estimated at 20¢/ton of coal processed (18). Among the more common and successful methods employed for phenol recovery from concentrated wastes, all are based upon extractive recovery into an immiscible organic solvent. Efficiencies are extremely high, with recoveries of 98–99% reported. However, even such high percent efficiencies can leave significant residual phenol in the treated effluent of a concentrated phenolic waste stream. The problem of solvent loss due to slight solubility in the wastewater versus efficiency of phenol removal must be balanced economically, as solvent costs can be high. In addition, any solvent dissolved in the extracted water represents increased organic matter in the waste. Recovery processes are summarized in Table 81, accompanied by reported operating data.

Table 81. Chemical Recovery Processes for Concentrated Phenolic Wastewaters (9).

Process	Influent Phenol (mg/l)	Effluent Phenol (mg/l)
Benzene-Caustic Dephenolization Process	3000	210-240
Countercurrent Podbielniak Extractors	2000	100
Pulsed Column Extractors	2200	30
Phenosolvan Dephenolization	1570	4.5
(Luigi)	2465	9.6
IFAWOL Dephenolization Process (Carl Still)	4000	40

The Koppers light oil extraction process has been reported to reduce 1500–2000 mg/l phenol to 10–30 mg/l (16). Other processes are also available (35). All extractive phenol processes appear capable of high recovery, but residual phenol may remain too high to meet effluent standards. It has been reported that phenol solutions of 7000 mg/l could be incinerated instead, for fuel costs of 1.5¢/1000 gal (36). Capital costs for the incinerator were estimated at $10,000–15,000.

Successful treatment of concentrated phenolic waste by activated carbon has been reported (37). The waste, initially at 2235 mg/l phenol, was reduced to 1950 mg/l by chemical coagulation and filtration. Activated carbon treatment yielded an effluent phenol level below 0.1 mg/l.

General cost data for concentrated waste treatment systems are summarized in Table 82. The data presented for the Barrett Phenol Recovery Process (35) are based on treatment to 5 mg/l. Capital

Table 82. Summary of Costs for Nonbiological Treatment Methods for High Concentration Phenol-Containing Wastewaters.

Method	Concentration (mg/l)	Flow (MGD)	Capital Cost ($/1000 gpd)	Treatment Cost (¢/1000 gal)	Reference
Incineration	7,000	—	10,000-15,000[a]	1.5	36
Extractor	2-4,000	0.15	3,333 (1969)	—	38
Podbielniak Extractor	11	0.2	2,500-3,000 (1969)	—	38
Barrett Phenol Recovery Process	5,000	0.01	9,000 (1957)		35
		0.10	2,000 (1957)		
		0.17	1,529 (1957)		
		0.50	1,340 (1957)		

[a] Total price, capacity not reported.

costs would decrease accordingly for lower levels. For example, if the final phenol concentration were 50 mg/l, capital costs for the 0.01- and 0.1-MGD plants would be $7000 and $1500/1000 gpd respectively. The 0.01-MGD plant costs reduce to $6000/1000 gpd for 90% phenol removal. Counterbalancing this gain, however, is decreased phenol recovery value.

Intermediate Level Treatment

An arbitrary classification for intermediate phenol concentrations might be 5–500 mg/l, since this encompasses the levels resulting from recovery process residuals up to levels that are not sufficiently high to warrant phenol recovery (from an economic viewpoint). In the absence of high concentrations of toxic substances or in the event of their successful prior removal, biological treatment is widely employed for treatment of wastewaters containing intermediate phenol levels. The biological processes include lagoons, oxidation ditches, trickling filters, and activated sludge. Table 83 summarizes reported performances of various biological processes employed for phenol removal. The activated sludge process is reported preferable, because of its high treatment efficiency and ease of control (24).

Phenol concentrations from 50 to 500 mg/l are generally considered suitable for treatment by biological processes, but as illustrated in Table 83, much higher levels have been successfully subjected to biological treatment. The exact concentration of phenol that can be directly treated biologically will depend to some extent upon other contaminants in the waste, and whether these contaminants are toxic to the microorganisms employed in the biological process. The cumulative effect of toxic constituents in phenolic wastes has been discussed by Reid and Libby, wherein chromium

Table 83. Performance of Biological Systems in the Treatment of Phenolic Wastes.

Treatment Process	Source of Industrial Waste	Phenol (mg/l) Influent	Phenol (mg/l) Effluent	Reference
Biooxidation pond	Refinery wastewater	30	1	27
Aerated lagoon	Municipal plus herbicide manufacturing waste	0.8-1.65	0.2-0.45	32
	(Same—Phenoxyacids)	2.15-4.14	1.51-1.73	32
Stabilization pond	Effluent from aerated lagoon above—phenols	0.2-0.45	0.06-0.13	32
	(Same—Phenoxyacids)	1.51-1.73	1.10-1.48	32
Oxidation ditch	Coke oven	990 (ave)	8.5 (ave)	19
Trickling filter	Oil refinery	30-40	0.5-0.7	21
	Oil refinery	—	.04	39
	Oil refinery	9-25	0.6-3.8	22
	Sourwater and plant sewage	15 (ave)	1.8 (ave)	
	Aircraft maintenance wastes	8	0.02	2
Activated sludge	Coke oven	400-700	10	20
	Coke oven	6.5	0.008	40
	Coke oven	9-10	0.9	9
	Coke oven	890	0.65	21
	Coke oven	110-200	0-11	10
	Coke oven with dephenolization	135	7	10
	Coke oven without dephenolization	1,400	15	10
	Oil refinery	80	<0.5	24
	Gas liquor	1,200	12	41
	Gas liquor	200	0.2	41
	Stocking manufacture	183-455	0.2-1.3	11
	Oil refinery	40-80	<1.0	23
	Fiberglass manufacturing	40-400	0.1-2.6	13
	Refinery and petrochemical	10-100	<1	29
Activated sludge plus Dow-Pac Filter	Oil refinery	40-80	0.35	23
Activated sludge[a]	Coke oven	3,350-3,900	0.2-0.8	15
Aero accelerator (activated sludge)	Coke oven	150	<1.0	42
Nocardia Process[b]	Foundary wastes	3,000	30-150	43
	Explosives	10,000	100	10

[a] Pilot plant—full scale plant in operation with lower loadings consistently produces 0.1 mg/l or 99.9% removal.

[b] Activated sludge with specially adapted organism culture for phenol removal.

accumulation gradually decreased the biological activity in a phenol treatment system (3).

Table 84. Treatment Costs Reported for Biological Systems Treating Phenolic Wastewaters.

Method	Flow (MGD)	Capital Cost ($/1000 gpd)	Treatment Cost (¢/1000 gal)	Reference
Activated sludge	0.11	2818 (1962)[a]	—	4
Activated sludge	1.92	417 (1965)	22	15
Activated sludge—sand filtration	2.88	868 (1970)	22	26
Air flotation—holding pond—trickling filter	6.5	154 (1968)	—	29

[a] year of cost data

Cost data reported for biological systems treating phenolic wastewaters are presented in Table 84. These values correspond closely to conventional biological unit operation costs. Generalized capital and operating costs of activated sludge and trickling filtration processes, as a function of design flow, are presented in Figure 37.

Other treatment methods have been employed for phenol removal at intermediate phenol concentrations. One of these is activated carbon, which seems to be competitive with biological treatment based on the data reported. These figures are presented in Table 85. Generalized cost data for activated carbon are presented in Figure 38. Although slightly higher in cost than biological systems, the advantage of consistency and immunity to reduced efficiency resulting from shock loading exists. Table 86 summarizes treatment results achieved with activated carbon.

Table 85. Treatment Costs for Activated Carbon Treatment of Phenolic Wastewaters.

Phenol (mg/l)	Flow (MGD)	Capital Costs ($/1000 gpd)	Treatment Costs (¢/1000 gal)	Reference
210	0.15	2000	36	31
370	0.5	2600	157	34
6	2.88	694	18	26

Note: All units include automatic carbon regeneration.

Several applications of chemical oxidation of phenols have been reported (8,46,47). Coagulation with alum and iron salts at various pH values removed only 10–20% of phenol at initial levels of

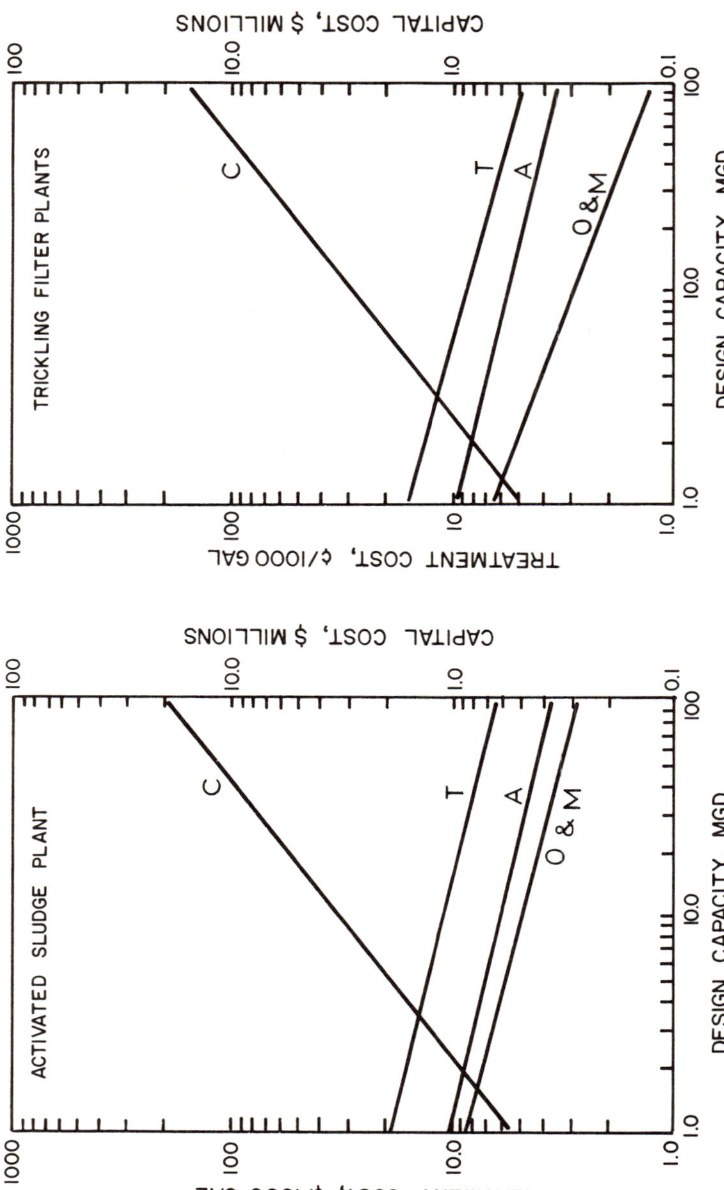

Figure 37. Costs for biological waste treatment systems. C = capital costs, $ millions; A = debt service (4.5% — 25 yrs), ¢/1000 gal; O & M = Operating and maintenance, ¢/1000 gal; T = total cost, ¢/1000 gal. [From Smith (44), courtesy Water Pollution Control Federation.]

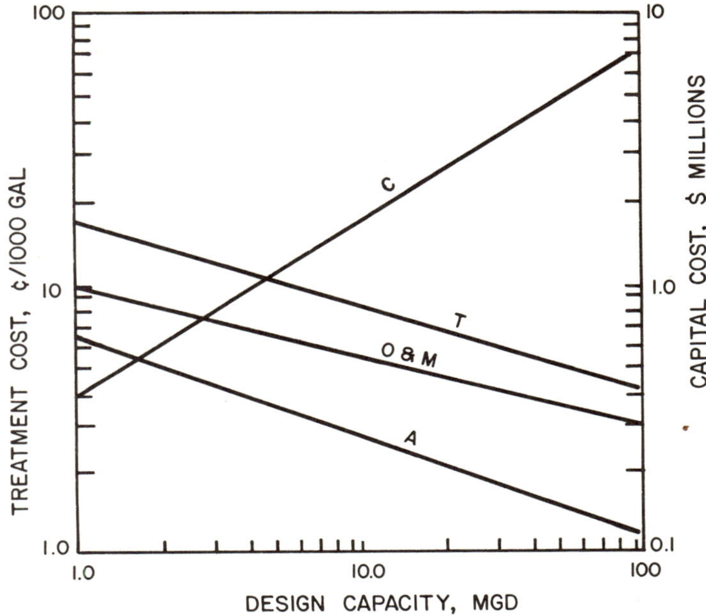

Figure 38. Costs of waste treatment by activated carbon. [From Smith (44), courtesy of Water Pollution Control Federation.]

Table 86. Activated Carbon Treatment of Phenolic Wastewaters (45).

Source	Phenol (mg/l) Initial	Final	Percent Reduction
Chemical manufacture	36.0	0.001	99+
Dyes	6.0	0.01	99+
Organic chemicals	0.52	0.05	99+
	0.12	0.003	99+
	0.315	0.001	99+
	12.8	0.001	99+
Inorganic chemicals	5325.0	0.25	99+
Plastics	9.75	0.013	99+
	8.5	0.006	99+
	26.5	0.005	99+
Explosives	16.6	0.023	99+
Petroleum refining	44.0	0.001	99+

100–125 mg/l (47). Only 62.4% removal was accomplished at 125 mg/l with permanganate oxidation, and at a cost of $1.83/1000 gal. However, it has been suggested that chlorination plus lime addition would result in 100% removal, at a cost of \$1.25–\$1.50/1000 gal (1952 prices). With chlorination, extremely high concentrations of chlorine must be added to effect complete phenol removal (46,47). Figure 39 presents treatment data for chlorination of phenolic wastes. Chlorination must be carried to completion and at pH less than 7, or toxic chlorophenols are produced (45). Using chlorination, no residual phenol, reduced from approximately 5 mg/l, resulted for rubber reclamation wastes (8). Treatment costs were estimated at \$1.00–\$1.66/1000 gal (1951 prices).

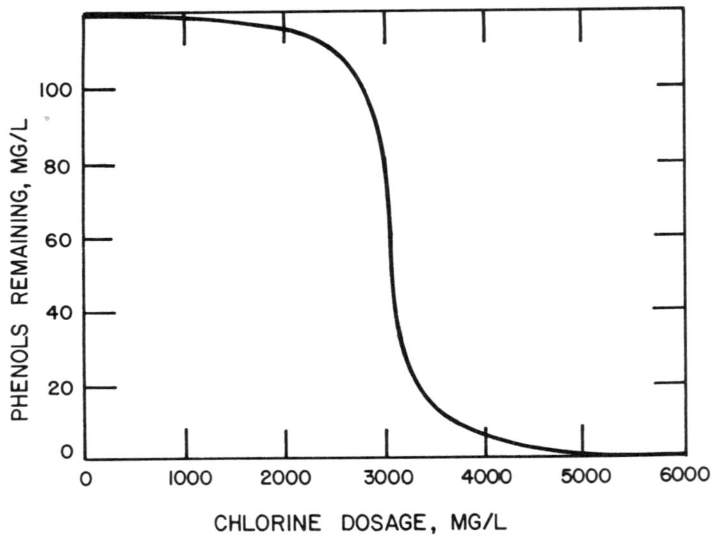

Figure 39. Phenol reduction by chlorine treatment. [From Cleary and Kinney (46), courtesy of Purdue University Industrial Waste Conference.]

Ozone and chlorine dioxide have also been reported as successful oxidants for phenol removal (8,10,12,20,21,23,24,46). The stoichiometric relationships between oxidant dose and phenol removals are shown in Figures 40 and 41 respectively (46). Although ozone is capable of efficient phenol removal, operating and capital costs may be extremely high, and ozone would likely be considered only as a low level or polishing treatment. Besselievre (38) reports that 1.5–2.5 parts ozone to 1 part phenol is effective for phenol removals down to 3 parts per billion.

A wastewater hydrocarbon stripper reduced initial phenol levels of 5–6 mg/l to less than 1 mg/l (48). Stripped gases were combusted

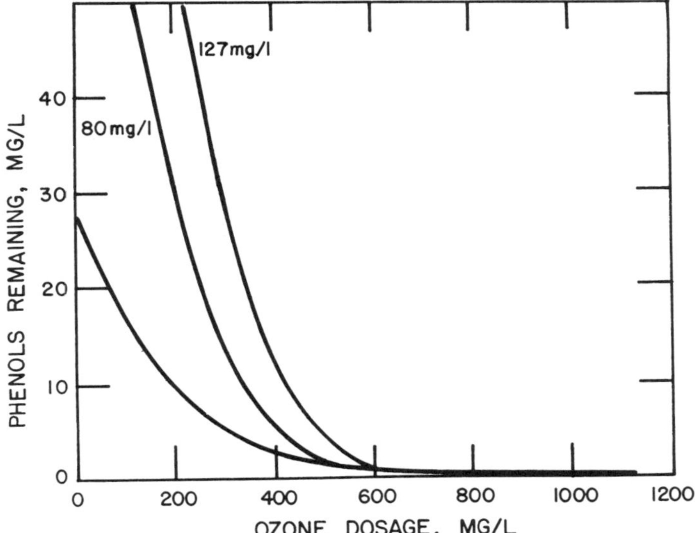

Figure 40. Phenol reduction by ozonation. Initial phenol concentrations as indicated. [From Cleary and Kinney (46), courtesy of Purdue University Industrial Waste Conference.]

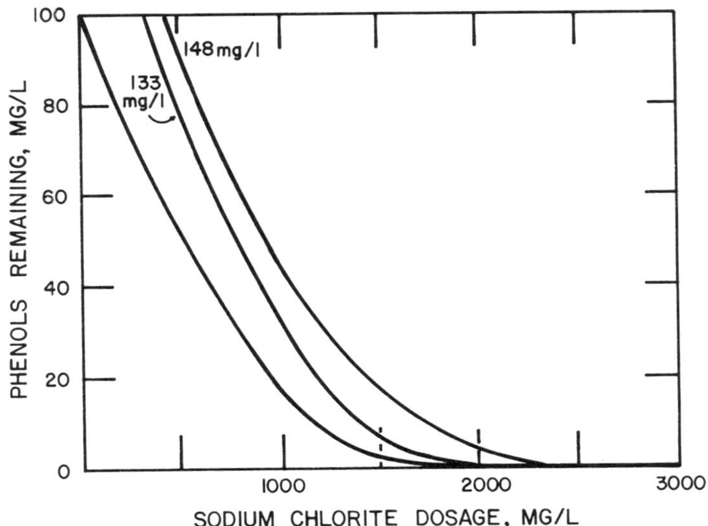

Figure 41. Phenol reduction by application of chlorine dioxide (as sodium chlorite). Initial phenol concentrations as indicated [From Cleary and Kinney (46), courtesy of Purdue University Industrial Waste Conference.]

rather than discharged to the atmosphere. Capital costs were estimated at $139/1000 gpd for a 0.18-MGD wastewater flow, and $74/1000 gpd for a 1.1-MGD wastewater flow. Operating costs for the fully automated units were estimated at 9.3¢ and 7.0¢/1000 gal respectively. Costs were expected to drop to 5–6¢/1000 gal for 1.3–2.0-MGD plants.

Dilute Wastes Treatment

Biological treatment generally is capable of reduction of phenol down to the 0.5–1 mg/l level, except for very high influent concentrations. Removal below the 1–0.1 mg/l level can be achieved by other processes. Normally, chemical or physical-chemical methods replace the biological processes for treatment of these dilute phenolic wastes.

Biologically treated refinery effluent with a phenol residual of 0.16–0.35 mg/l was reduced to 0.003 mg/l with ozone treatment (23). The ozone treatment was reportedly so effective that an activated carbon polishing unit designed to follow the ozonator was not required.

Activated carbon use is also well established for removal of trace organics as a tertiary treatment process, and is equally effective for moderate strengths and dilute phenolic wastes, as shown in Table 86. Activated carbon has been reported to have a phenol capacity of 0.5 lb/100 lb carbon, up to 25 lb/100 lb (23,45). The process is most effective at lower pH for phenol and other weak organic acids.

Recycle Practice

Recycle of phenolic wastewaters, with or without phenol removal has been reported. Several reports (12-14) have detailed the reuse of rinse water to wash continuous formation belts in fiberglass manufacturing. The water cycles in the rinse system, and is used as formation water for the phenolic resin binder sprayed onto the glass. Therefore, soluble phenolics are recovered and reused. Blowdown of 20–40% is necessary to prevent excessive solids buildup. Recycle water characteristically contains 240 mg/l phenol, 16,000 mg/l dissolved solids and 200 mg/l suspended solids (13). This would indicate that treatment is mandatory for the blowdown, since total phenol levels are substantial.

Savings in phenolic resin costs of $2700/mo (12) to $40,000/mo (13) have been reported. A cost analysis for such a system (14) reported a reduction in water costs from 75¢/1000 gal (municipal charges plus 12¢/1000 gal in-plant-softening costs) to a net cost of 37¢/1000 gal. Total capital and gross operating costs for the recovery system were reported as 159¢/1000 gal. Capital investment is approximately $1700/1000 gpd for 24 hr/day operation at 0.11 MGD.

One refinery (28) has incorporated biological treatment directly into its recycle system. Phenol wastewaters are employed as makeup to the cooling water system at 800–1000 gpm. The cooling towers function as biological reactors for both forced draft and induced draft towers. The blowdown of 200–500 gpm typically contains 0.082–9.80 mg/l of phenol, representing 99.4–99.9+% removal. Nine years' operating experience has substantiated that limited clogging and corrosion problems exist.

As early as 1953, a synthetic phenol production facility reduced its annual water use from 17,427,000 gal/yr containing an average phenol of nearly 8000 mg/l to 783,000 gal containing no phenols. Abatement costs were given as $100,000 but chemical savings of $112,000/yr were realized (49).

SUMMARY

Phenol removal to very low residual levels is technologically feasible and in practice. Removals to a few parts per billion are accomplished by multiple treatment sequences, since increasing efficiency for any one process usually is associated with a substantial increase in cost. Processes are available for 99% or greater removals for the 500–10,000 mg/l phenol level. Such processes are typically product recovery oriented. Biological processes have been shown capable of up to 99% phenol removal in the 10–500 mg/l range. Specialized biological systems are capable of treating even higher concentrations. Although not favorable for direct treatment of concentrated or intermediate level phenolic wastes, chemical oxidation processes such as chlorination and ozonation demonstrate favorable economics when employed as polishing processes after biological systems.

Carbon adsorption is in use for both intermediate and dilute phenol levels, and can display a wide range of capital and operating costs depending upon the specificity of the contaminant removal. Water reuse, both with and without phenol degradation has been reported successful.

Costs associated with treatment processes generally are of the order of equivalent water and wastewater treatment processes under municipal management.

REFERENCES

1. McKee, J. E. and H. W. Wolf. *Water Quality Criteria,* 2nd ed., California State Water Quality Control Board, Publication No. 3-A (1963).
2. Reid, G. W., R. Daigh and R. L. Wortman. "Phenolic Wastes from Aircraft Maintenance," *J. Water Poll. Cont. Fed.* **32,** 353-391 (1960).
3. Reid, G. W. and R. W. Libby. "Phenolic Waste Treatment Studies," *Proc. 12th Purdue Ind. Waste Conf.* **12,** 250-258 (1957).
4. Barzler, R. P., D. J. Giffels and E. Willoughby. "Pollution Control in Foundry Operations," In *Industrial Pollution Control Handbook,* Herbert F. Lund, Ed. (New York: McGraw-Hill Book Co., 1971).
5. Schesinger, H. A., E. F. Dul and T. A. Fridy, Jr. "Pollution Control in Textile Mills," In *Industrial Pollution Control Handbook,* Herbert F. Lund, Ed. (New York: McGraw-Hill Book Co., 1971).
6. Ross, R. D. *Industrial Waste Disposal* (New York: Van Nostrand Reinhold Co., 1968).
7. Parsons, W. A. *Chemical Treatment of Sewage and Industrial Wastes,* (Washington, D.C.: National Lime Association, 1965).
8. Sechrist, W. D. and N. S. Chamberlin. "Chlorination of Phenol Bearing Rubber Wastes," *Proc. 6th Purdue Ind. Waste Conf.* **6,** 396-412 (1951).
9. Wurm, H. J. "The Treatment of Phenolic Wastes," *Proc. 23rd Purdue Ind. Waste Conf.* **23,** 1054-1073 (1968).
10. Noack, Werner (Formal Discussion to Biszysko and Suschka). "Investigations on Phenolic Wastes Treatment in an Oxidation Ditch," In *Advances in Water Pollution Research, Munich, Conference,* Vol. 2 (Elmsford, New York: Pergamon Press, Inc., 1967), pp. 285-295.
11. Ide, T. (Formal Discussion to Biczysko and Suschka). "Investigations on Phenolic Wastes Treatment in an Oxidation Ditch," In *Advances in Water Pollution Research, Munich Conference,* Vol. 2 (Elmsford, New York: Pergamon Press, Inc., 1967), pp. 285-295.
12. Baloga, J. M., F. B. Hutto, Jr. and E. I. Merrill. "A Solution to the Phenolic Problem in Fiberglass Plants," *Water Sewage Works* **118,** 7-13, (1971).
13. Fletcher, G. W., S. H. Thomas and D. E. Cross. "Development and Operation of a Closed Wastewater System for the Fiberglass Industry," Presented at the 45th Annual Water Poll. Cont. Fed. Meeting (1972).
14. Johns-Manville Products Corporation. "Phenolic Waste Reuse by Diatomite Filtration, U.S. EPA Report 12080EZF 09/70 (1970).
15. Kostenbader, P. D. and J. W. Flecksteiner. "Biological Oxidation of Coke Plant Weak Liquor," *J. Water Poll. Cont. Fed.* **41,** 199-207 (1969).
16. Fisher, C. W. "Coke and Gas," In *Chemical Technology Volume 2, Industrial Wastewater Control,* F. Fred Gurnham, Ed. (New York: Academic Press, Inc., 1965).

17. Carbone, W. E., R. N. Hall, H. R. Kaiser and C. G. Bazell. "Commercial Dephenolization of Ammonical Liquors with Centrifugal Extractors," *Proc. 5th Ontario Ind. Waste Conf.* 42-58 (1958).
18. Resource Engineering Associates. *State of the Art Review on Product Recovery* (Washington, D.C.: U.S. Dept. of the Interior, 1969).
19. Biszysko, J. and J. Suschka. "Investigations of Phenolic Wastes Treatment in an Oxidation Ditch," In *Advances in Water Pollution Research, Munich Conference*, Vol. 2 (Elmsford, New York: Pergamon Press, Inc., 1967).
20. Clough, G. F. S. "Biological Oxidation of Phenolic Waste Liquor," *Chem. Proc. Eng.* **42** (1), 11-14 (1961).
21. Lesperance, T. W. "Biological Treatment of Phenols," *Proc. 8th Ontario Ind. Waste Conf.* 59-66 (1961).
22. Graves, B. S. "Biological Oxidation of Phenols in a Trickling Filter," *Proc. 14th Purdue Ind. Waste Conf.* **14,** 1-6 (1959).
23. McPhee, W. T. and A. R. Smith. "From Refinery Waste to Pure Water," *Proc. 16th Purdue Ind. Waste Conf.* **16,** 311-326 (1961).
24. Benger, M. "The Disposal of Liquid and Solid Effluents from Oil Refineries," *Proc. 21st Purdue Ind. Waste Conf.* **21,** 759-767 (1966).
25. Steck, W. "The Treatment of Refinery Waste Water with Particular Consideration of Phenolic Streams," *Proc. 21th Purdue Ind. Waste Conf.* **21,** 783-790 (1966).
26. Peoples, R. F., P. Krishnan and R. N. Simonsen. "Nonbiological Treatment of Refinery Wastewater," *J. Water Poll. Cont. Fed.* **44,** 2120-2128 (1972).
27. Wigren, A. A. and F. L. Burton. "Refinery Wastewater Control," *J. Water Poll. Cont. Fed.* **44,** 117-128 (1971).
28. Mohler, E. F., Jr., H. F. Elkin and L. R. Kumnick. "Experience with Reuse and Biooxidation of Refinery Wastewater in Cooling Tower Systems," *J. Water Poll. Cont. Fed.* **36,** 1380-1392 (1964).
29. Carnes, B. A., J. M. Eller and J. C. Martin. "Reuse of Refinery and Petrochemical Wastewaters," *Ind. Water Eng.* **9,** 25-29 (1972).
30. Dickenson, B. W. and W. T. Laffey. "Pilot Plant Studies of Phenol Waste from Petrochemical Operations," *Proc. 14th Purdue Ind. Waste Conf.* **14,** 780-799 (1959).
31. Henshaw, T. B. "Adsorption/Filtration Plant Cuts Phenols from Effluent," *Chem. Eng.* **78,** 47-49 (1971).
32. Sidwell, A. E. "Biological Treatment of Chlorophenolic Wastes," U.S. EPA Report 12130EGK 06/71 (1971).
33. *The Cost of Clean Water, Vol. III, Industrial Waste Profile No. 10, Plastics Materials and Resins* (Washington, D.C.: U.S. Dept. of the Interior, 1967).
34. Schumaker, T. P. and R. H. Zanitsch. "Physical/Chemical Treatment: A Solution to a Complex Waste Problem," presented at the 45th Annual Water Poll. Cont. Fed. Meeting (1972).
35. Heller, A. N., E. W. Clark and W. Reiter. "Some Factors in the Selection of a Phenol Recovery Process," *Proc. 12th Purdue Ind. Waste Conf.* **12,** 103-122 (1957).
36. Ross, R. D. "Pollution Waste Control," In *Industrial Pollution Control Handbook*, Herbert F. Lund, Ed. (New York: McGraw-Hill Book Co., 1971).

37. Van Stone, G. R. "Treatment of Coke Plant Effluent," *Ind. Waste* **18** (4), 23-35 (1972).
38. Besselievre, E. B. *The Treatment of Industrial Wastes* (New York: McGraw-Hill Co., 1969).
39. Sample, G. E. and R. D. Rea. "Floats Away Refinery Wastes, Oil, Phenols Reduced 98%," *Chem. Proc.* **32** (12), 41-42 (1969).
40. Pinkstaff, E., C. A. Leahu and G. F. Gurnham. "Phenol Removal from Coke Plant and Petroleum Refinery Wastewaters by Municipal Treatment," *Water Poll. Abstr.* **40,** No. 632 (1967).
41. Nakashio, M. "Phenolic Wastes Treatment by Activated Sludge Process. The Operation Condition and Actual Operation," *J. Ferment, Technol. (Osaka)* **47,** 389-393 (1969); *Water Poll. Abstr.* **43,** No. 1025 (1970).
42. Heinicke, D. "Developments in Reclamation as Applied to Chemical Effluents," In *Disposal of Process Wastes Liquids, Solids, Gases. A Symposium,* Max Wulfinghoff, Trans. (New York: Chemical Publishing Co., Inc., 1968).
43. Schweisfurth, R. and G. Schertz. "Investigations on Biological Treatment of Ammonical Foundry Wastewater by the Nocardia Process," *Gesundheitsing* **83,** 273 (1962); *Water Poll. Abstr.* **37,** No. 139 (1964).
44. Smith, R. "Cost of Conventional and Advanced Treatment of Wastewater," *J .Water Poll. Cont. Fed.* **40,** 1546-1547 (1968).
45. Hager, D. G. "Industrial Wastewater Treatment by Granular Activated Carbon," *Ind. Water Eng.* 14-28 (January-February, 1974).
46. Cleary, E. J. and J. E. Kinney. "Findings from a Cooperative Study of Phenol Waste Treatment," *Proc. 6th Purdue Ind. Waste Conf.* **6,** 158-170 (1951).
47. Chamberlin, N. S. and A. E. Griffin. "Chemical Oxidation of Phenolic Wastes with Chlorine," *Sewage Ind. Wastes* **24,** 750-760 (1952).
48. Schutt, H. C. and J. Loftus. "Waste Water Conditioned by Carrier Gas," *Oil Gas J.* **64** (32), 70-72 (1966).
49. Heller, A. N. "The Organic Chemical Industry," *Sewage Ind. Wastes* **28,** 665-671 (1956).

19

TREATMENT TECHNOLOGY FOR SELENIUM

Industries using selenium include paint, pigment and dye producers, electronics, glass manufacturers, and insecticide industries. Essentially no information is available in the water pollution literature on levels of selenium in industrial wastewaters, treatment methods for selenium wastes, or costs associated with removal of selenium from industrial wastewaters.

It has been reported that selenium is present in almost all types of paper, and it might be concluded from this information that pulp and paper mill wastes could contain selenium (1). One primary concern has been the release of selenium as an air pollutant, upon incineration of paper products (1). Selenium wastewater levels have been reported as follows. For incinerator fly ash quench water, soluble selenium was measured at 5–23 μg/l, while incinerator residue quench water contained 3 μg/l of soluble selenium. The incinerator handled a solid waste containing approximately 55–69% paper (1).

Selenium occurs as an impurity in the form of selenide, Se^{-2}, in metallic sulfide ores (2). When the sulfide ores are roasted in air (to convert the sulfide and drive off SO_2) the selenide is oxidized to selenium dioxide and is released in the flue gas. If the flue gas is quenched, selenium dioxide reacts with the quench water to form selenious acid, H_2SeO_3. A selenium level of 0.7 mg/l has been reported in wastewaters from copper smelting and electrolytic refining (3). The selenite ion appears to be the most common form of selenium in wastewater except for pigment and dye wastes, which contain the selenide (*e.g.*, yellow cadmium selenide). Other forms decompose to yield the SeO_3^{-2} ion. It has also been reported that selenium dioxide is readily reduced, to precipitate finely divided elementary selenium. It is probable that this same reaction might occur for selenious ion at acidic pH. Selenide may be precipitated as the metallic salt, which is high insoluble (2). Raible (4) noted the presence of selenium in a pharmaceutical wastewater, but did not indicate concentrations.

Secondary municipal sewage treatment plant effluents containing 2–9 μg/l of selenium have been reported (5). A tertiary sequence of treatment, which included lime treatment to pH 11, sedimentation, mixed media filtration, activated carbon adsorption and chlorination yielded selenium removal of 0–89%. Best removal was obtained at higher initial selenium concentrations. Activated carbon treatment of a secondary treated municipal wastewater reduced selenium from 9.32–5.85 μg/l. This represents a 37% removal efficiency (6). Logsden *et al.*, however, have reported less than 4% removal of selenite ion by activated carbon, from well water (7). This suggests that the enhanced removal observed in treated municipal sewage (6) might be more influenced by wastewater organic content than by inorganic selenite adsorption. Results of pilot studies on selenite removal are presented in Table 87. Although neither ferric sulfate nor alum coagulation was very effective, treatment improved for both coagulants with increasing coagulant dosage and decreasing pH.

Linstedt, *et al.* (8) measured 2.3 μg/l of selenium in the effluent from a secondary sewage treatment plant. These workers investigated the efficiencies of removal of selenium, as SeO_3^{-2}, by several advanced wastewater treatment processes and reported the results shown in Table 88. The authors concluded that efficient removal of selenium could be achieved with a strong acid-weak base ion exchange system. Figure 42 presents capital and operating costs for such a weak-base system, as a function of dissolved solids content of the wastewater. The nonselective nature of ion exchange resins is a drawback when attempting to remove a low-level contaminant in the presence of significant quantities of other ions. The simultaneous removal of other ions rapidly increases ion exchange costs. Other commonly employed industrial metal treatment processes (*e.g.*, lime coagulation, settling, and sand filtration) were ineffective in recovering selenium, at least as the negatively charged anion. In that event, ion exchange appears to be the most effective technique for which actual results have been reported.

Selenium treatment by precipitation upon addition of a sulfide salt at slightly acidic (6.5) pH has been suggested (10). Although no treatment results were cited, this treatment technique is claimed to yield selenium effluent levels of 0.05 mg/l. The likely treatment mechanism involved is reduction of the selenite ion, precipitating elemental selenium. Sulfide would be cooxidized in the process.

Table 87. Results of Pilot Treatment Studies on Selenite[a] (7).

Treatment and Dosage	pH	Percent Removal
Ferric sulfate coagulation		
25 mg/l	7.0	60
	5.5	78
60 mg/l	7.0	75
	5.5	83
100 mg/l	7.0	81
	6.0	84
Alum coagulation		
25 mg/l	6.0	10
100 mg/l	6.0	27
200 mg/l	6.0	45

[a] Initial selenium concentration = 0.09-0.10 mg/l.

Table 88. Removal of Selenium by Bench-Scale Advanced Wastewater Treatment Processes (8).

Process	Removal (Percent)
Lime coagulation-settling	16.2
Cation exchange	0.9
Cation plus anion exchange	99.7
Process sequence[a]	
1st Sand filtration	9.5
2nd Activated carbon	43.2
3rd Cation exchange	44.7
4th Anion exchange	99.9

[a] Cumulative removal after indicated process.

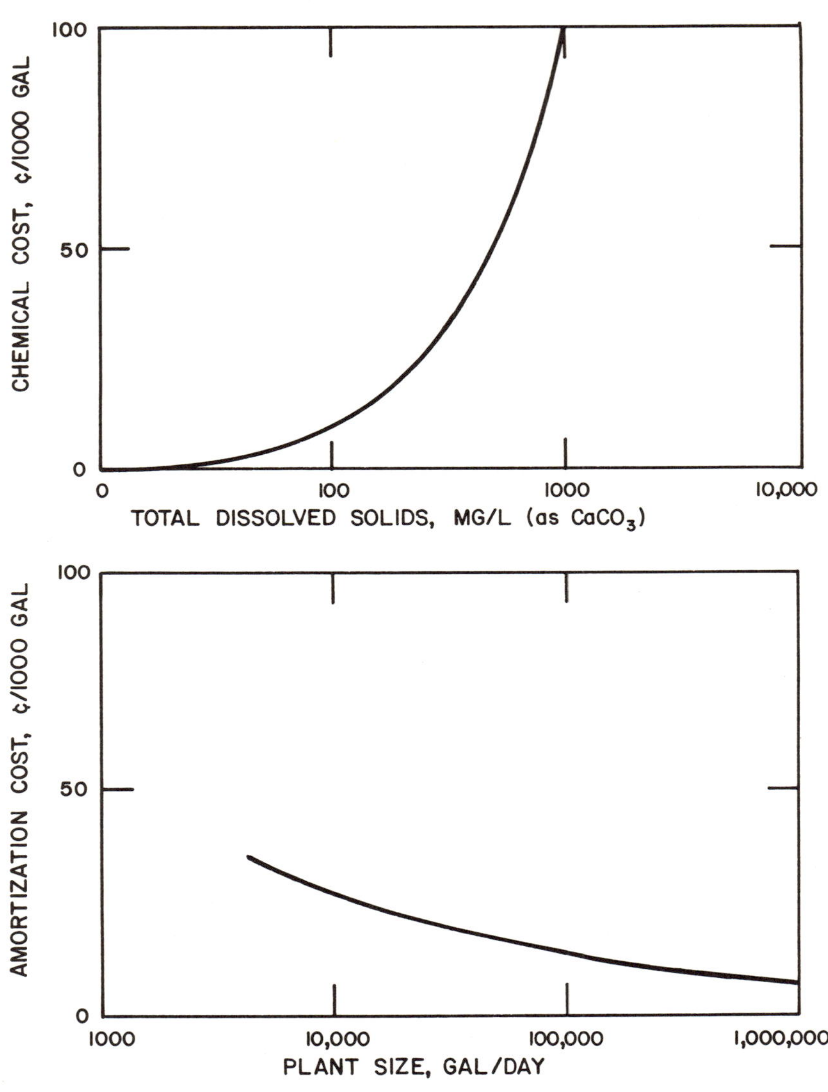

Figure 42. Chemical and amortization costs for a weak-base ion exchange system. [From Ahlgren (9), courtesy of *Industrial Water Engineering.*]

REFERENCES

1. Johnson, H. "Determination of Selenium in Solid Waste," *Environ. Sci. Technol.* **4,** 850-853 (1970).
2. Hutchinson, E. *Chemistry: The Elements and Their Reactions* (Philadelphia, Pennsylvania: W. B. Saunders Co., 1959).
3. Hallowell, J. B., J. F. Shea, G. R. Smithson, Jr., A. B. Tripler and B. W. Gearser. "Water-Pollution Control in the Primary Nonferrous Metals Industry, Volume I, Copper, Zinc and Lead Industries," U.S. EPA Report EPA-R2-73-247a (September, 1973).
4. Raible, J. A. "Fluorometric Determination of Selenium in Effluent Streams with 2, 3-Diaminonaphthalene," *Environ. Sci. Technol.* **6,** 621-622 (1972).
5. Argo, D. G. and G. L. Culp. "Heavy Metals Removal in Wastewater Treatment Processes: Part 2—Pilot Plant Operation," *Water Sewage Works* **119,** 28-132 (1972).
6. Culp, G. L. and R. L. Culp. *New Concepts in Water Purification* (New York: Van Nostrand Reinhold Co., 1974).
7. Logsden, G. S., T. J. Serg and J. M. Symons. "Removal of Heavy Metals by Conventional Treatment," *Proc. 16th Water Qual. Conf.*, University of Illinois (February 12-13, 1974).
8. Linstedt, K. D., C. P. Houch and J. T. O'Connor. "Trace Element Removals in Advanced Wastewater Treatment Processes," *J. Water Poll. Cont. Fed.* **43,** 1507-1513 (1971).
9. Ahlgren, R. M. "Membrane vs. Resinous Ion Exchange Demineralization," *Ind. Water Eng.* **8,** 12-14 (1969).
10. Curry, N. A. "Philosophy and Methodology of Metallic Waste Treatment," Presented at 27th Ind. Waste Conf., Purdue University (1972).

20

TREATMENT TECHNOLOGY FOR SILVER

MAJOR INDUSTRIAL SOURCES

Silver, as the solid metal, is used in the jewelry, silverware, metal alloy, and food and beverage processing industries. Little soluble silver waste would be expected to result from use of the solid metal. The only appreciably soluble common silver salt—silver nitrate—is used in the porcelain, photographic, electroplating, and ink manufacture industries, and also as an antiseptic (1). The two major sources of soluble silver wastes are the photographic and electroplating industries.

TREATMENT TECHNOLOGY

Recovery of silver from waste streams is more lucrative than for most other metals. Silver cyanide plating baths may contain from 13,000–45,000 mg/l of silver (2). This can represent appreciable silver loss by drag-out. About 75% of silver in the photographic industry finds its way into the fixing solution (3). A recent study on product recovery places a net recovery value of $1.60–$9.00/1000 gal of dilute plating rinse waters containing from 50–250 mg/l of silver (4).

Basic methods for silver removal from wastewaters are discussed under four categories: 1) precipitation, 2) ion exchange, 3) reductive exchange, and 4) electrolytic recovery. More than one of these types of processes may be combined, to achieve high levels of silver removal.

Precipitation

Silver is frequently removed from wastewaters by precipitation as silver chloride. By contrast to most metal chlorides, which are relatively soluble, silver chloride is an extremely insoluble salt. The salt will dissolve in water to a maximum concentration of approximately 1.4 mg/l silver ion (5). Slight excess of chloride ion will reduce this value, but greater excess increases the solubility of silver through the formation of soluble silver chloride complexes (5,6). Thus, silver can be selectively recovered as silver chloride from a mixed metal waste stream without initial waste-stream segregation or concurrent precipitation of other metals. If treatment conditions are alkaline, resulting in precipitation of hydroxides of other metals along with the silver chloride, acid washing of the precipitated sludge removes contaminant metal ions, leaving the insoluble silver chloride.

Plating wastes generally contain silver in the form of silver cyanide, which interferes with precipitation of silver chloride. Therefore, cyanide removal is necessary prior to precipitation of silver as the chloride salt. Oxidation of the cyanide with chlorine releases chloride ions into solution, which in turn react to form silver chloride directly (6). In the event that cyanide is present in great excess over silver, high chloride concentrations resulting from cyanide oxidation may greatly reduce the effectiveness of this silver removal process.

Table 89. Results of Bench-Scale Experiments on Silver Removal by Chlorination (6).

	Cyanide (mg/l)		Silver (mg/l)	
Experiment	Initial	Final	Initial	Final
1	102	0	105	3.5
4	208	0	250	2.8
10	156	0	150	1.0

Bench-scale studies, performed in designing a silver waste treatment process, yielded the results presented in Table 89. Treatment consisted of addition of chlorine in the ratio of 3.5 mg/mg of cyanide, and a 30-minute reaction time. Silver reductions of 90–99% were achieved. Modification of the batch treatment process to the sequence described below resulted in even higher removals, as indicated in Table 90:

a) Chlorination and 10-minute reaction
b) Adjust pH to 6.5 to complete oxidation of cyanide
c) Addition of ferric chloride
d) Skimming and sedimentation
e) Lime addition to pH 8
f) Settling of floc and decanting supernatant.

Table 90. Results of Bench-Scale Experiments on Treatment of Plating Wastes (6).

	Cyanide (mg/l)		Silver (mg/l)		Copper (mg/l)		Nickel (mg/l)		Zinc (mg/l)	
No.	Initial	Final	Initial	Final	Initial	Final	Initial	Final	Initial	Final
1	9.4	0.35	0.7	0.0	18	0.5	0.1	0.0	0.3	0.2
2	4.2	0.3	0.8	0.0	14	0.1	0.1	0.0	4	0.7
3	5.2	0.2	12	0.0	17	0.05	0.8	0.1	10	1.1
4	5.4	0.2	25	—	18	17	13	0.2	8	4.8
5	25	—	8	0.0	4.5	1.5	2.3	1.0	9	0.5
6	4	—	40	0.0	16	2.3	10	0.9	11	2.4

Improved removal resulted, reducing the residual silver ion in the effluent to less than 0.1 mg/l. High removal of other heavy metals was achieved with both methods 1 and 2. The treatment plant resulting from the above studies handles silver recovery in batch

Table 91. Results of Operation of Batch Silver Recovery System During One Month (7).

	Raw Waste		Effluent	
Batch No.	Cyanide (mg/l)	Silver (mg/l)	Cyanide (mg/l)	Silver (mg/l)
1	642	564	70	8.2
2	829	585	86	4.4
3	509	401	35	2.7
4	348	331	30	6.5
5	389	364	54	2.7
6	672	533	56	5.1
7	431	440	40	0.7
8	159	130	15	0.0

processes of 10,000 gal. Operating data from the full-scale recovery system is presented in Table 91.

Other silver precipitation processes are mentioned in the literature. One, a patented process of Eastman Kodak Co. (8), precipitates silver from waste photographic solutions containing high levels of organic acids by adding magnesium sulfate and lime. The silver likely precipitates as a mixed sulfate-oxide and is recovered from the sludge.

Precipitation of silver by joint lime-ozone treatment has been examined in bench-scale studies (9). The initial silver concentration of 91 mg/l was reduced to the following levels after treatment:

	Residual Soluble Silver (mg/l)	
pH	Lime Only	Lime + Ozone
9	37	1.4
10	12	0.1
11	0.4	0.1

Use of sulfide for precipitating silver from photographic solutions as the extremely insoluble silver sulfide is also reported (10), but no operating data or costs are provided. From a solubility standpoint, sulfide precipitation would appear to be an attractive method, although solids separation and addition of sulfide to the wastewater are considered disadvantages. In a review of methods of silver recovery in the photographic film processing industry, Schreiber (11) has pointed out that the use of sulfide is one of the oldest methods for silver precipitation. Use of hydrosulfite for precipitation yields both free silver and the sulfide, as well as a more compact precipitate with good settling characteristics. Heat is required, however, and chemical costs are high.

Ion Exchange

A patented process involves the recovery of silver from dilute photographic wash waters by passage through basic ion exchange resins (12). Recovery is by elution of silver salts from the resin, or by direct resin incineration and recovery of pure silver. Above 85% removal of trace levels of silver were achieved from an extremely dilute secondary sewage effluent by cation exchange, and 91.7% removal by combined cation-anion exchange (13).

A series cation-anion exchanger combination has been used for recovery of silver cyanide from plating rinse waters (14). The recovered sodium silver cyanide may be returned to the plating bath, or converted to silver chloride by the precipitation process. Chemical costs of recovery are given as 18¢/lb of silver recovered.

A five-column ion exchange recovery system has been reported for silver cyanide rinse waters (15). The first column contains a cation exchange resin which converts the silver cyanide to a silver hydrogen cyanide complex. This in turn passes through the second column (an anion exchange resin) where silver cyanide is removed. The additional three columns function to effect further cyanide removal. The eluant from the second column is treated with potassium or sodium cyanide to recover the silver as potassium or sodium silver cyanide.

The cost of silver cyanide waste treatment by ion exchange is reported as $1.75/1000 gal (16). However, the value of silver recovered results in a net profit of $27.20/1000 gal of wastewater treated. Approximately 280 mg of silver were recovered from each liter of wastewater (16). Recovery costs would be somewhat higher if the silver chloride precipitation method of recovery was practiced, as cyanide destruction would be a prerequisite to formation of the silver chloride. Approximate capital costs of ion exchange units are given in Figure 43. These costs are for complete units, including valving and resins.

Problems in recovering silver from the resin may make ion exchange uneconomical for photographic waste treatment. Ion exchange in conjunction with electrolytic recovery for thiosulfate developer baths has been suggested (18). In that case, fixing solution silver is recovered by plating out the silver. Washout silver waste is accumulated by ion exchange, and the resin regenerant containing the concentrated silver is passed through an electrolytic process, thereby removing silver and functioning as developer bath makeup water.

The development of a silver specific ion exchange resin, similar to that available for gold recovery is reported underway (19). The application of the projected resin is for silver recovery from spent fixer solutions.

Reductive Exchange

This method involves cementation of silver onto a sacrificial metal. Usually zinc or iron is used because they are relatively inexpensive. The net result is a substitution of iron or zinc ion for

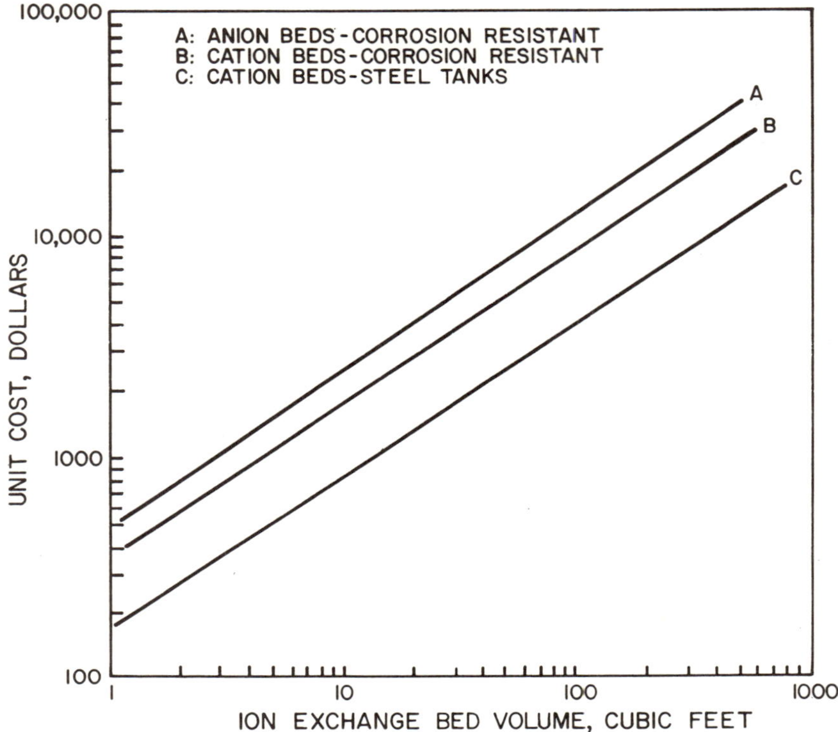

Figure 43. Costs of ion exchange units. [From McGarvey, *et al.* (17), courtesy of *Industrial Engineering Chemistry*.]

silver in solution. This process had been in operation at one plant prior to installation of a cyanide chlorination-silver chloride recovery system (20). Although the process was effective for silver recovery, replacement of the reductive exchange process was necessary for effective cyanide treatment.

Using zinc or steel wool under proper conditions to maintain sufficient metal surface exposed, 95% removal of silver can be accomplished with column operation cementation. Silver ions are replaced with equivalent quantities of zinc or iron ions.

The potential use of activated carbon as a treatment method for silver has been reported (21). The mechanism is described as one of reductive recovery by formation of elemental silver at the carbon surface. Laboratory results indicate that the carbon is capable of retaining silver to 9% of carbon weight at pH 2.1 and 12.5% at pH 5.4.

Electrolytic Recovery

This process has been reported in conjunction with photographic developer solutions (22). As with any electrolytic process, high wastewater ionic concentrations are necessary for the method to function effectively. If sufficient metallic ions are in solution such an operation is feasible; however, most silver wastewaters do not appear to contain sufficient concentrations of silver ion alone. Many electrolytic units will operate at silver concentrations of 500 mg/l, and some as low as 100 mg/l. Usually operation begins at 5000 mg/l, however, and reduces bath silver concentrations to 500 mg/l.

Closed-loop operation of the first rinse after silver plating baths is economically feasible for electrolytic recovery using laminar or packed bed recovery cells (23). High-electrode efficiency makes the process worthwhile.

SUMMARY

The value of silver makes recovery from process streams attractive. Electrolytic recovery, as long practiced in photographic processing, is operative only down to solution concentrations of 100–500 mg/l. However, packed bed electrodes may find use to much lower concentrations. Precipitation with chloride ion can remove silver to the milligram per liter level. Coprecipitation with other metal hydroxides under alkaline conditions improves silver removal to less than 0.1 mg/l. Silver recovery is possible from mixed precipitates by redissolving the metal precipitates under acidic conditions. Ion exchange for low silver concentrations appears feasible for separated waste streams, where total ionic strength is not appreciably greater than the silver salt concentration. Very low residual silver concentrations are possible with ion exchange. Sulfide precipitation of silver has been used on a small scale for many years, but large-scale operation appears to pose problems, particularly with respect to solids separation and handling.

REFERENCES

1. McKee, J. E. and H. W. Wolf. *Water Quality Criteria,* 2nd ed., California State Water Quality Control Board Publication No. 3A (1963).
2. Lowe, W. "The Origin and Characteristics of Toxic Wastes, with Particular Reference to the Metals Industry," *Water Poll. Cont. (London)* 270-280 (1970).

3. Kuhn, A. T. "Electrochemical Techniques for Effluent Treatment," *Chem. Ind.* 946-950 (1971).
4. Resource Engineering Associates. "State of the Art Review on Product Recovery," (Washington, D.C.: U.S. Dept. of the Interior, 1969).
5. Butler, J. N. *Ionic Equilibrium* (Reading, Massachusetts: Addison-Wesley Publishing Co., 1964).
6. Walker, C. A., W. Zabban, R. W. Southworth and E. P. Heslin, "Disposal of Electroplating Wastes by Oneida, Ltd. II. Development of Treatment Processes," *Sewage Ind. Wastes,* **26,** 849-853 (1954).
7. Eichenlaub, P. W. and J. Cox. "Disposal of Electroplating Wastes by Oneida, Ltd. V. Plant Operation," *Sewage Ind. Wastes* **26,** 1130-1135 (1954).
8. Pool, S. C. U.S. Patent 2,507,175, to Eastman Kodak Co., *Water Poll. Abstr.* **24,** No. 123 (1951).
9. Netzer, A., A. Bowers and J. D. Norman. "Removal of Trace Metals from Wastewater by Lime and Ozonation," In *Pollution: Engineering and Scientific Solutions* (in press).
10. Fusco, R. "Recovery of Silver from Photographic Fixing Baths and Other Sources," *Chem. Ind.* **24,** 356 (1942); *Water Poll. Abstr.* **21,** No. 389 (1948).
11. Schreiber, M. L. "Present Status of Silver Recovery in Motion-Picture Laboratories," *J. Soc. Motion Pict. Tech. Eng.* **74,** 505-513 (1965).
12. Permutit Co., Ltd., "Improvements Relating to the Recovery of Silver from Dilute Solutions," B. P. 626,081; *Water Poll. Abstr.* **23,** No. 623 (1950).
13. Linstedt, K. D., C. P. Houch and J. T. O'Connor. "Trace Element Removals in Advanced Wastewater Treatment Processes," *J. Water Poll. Cont. Fed.* **43,** 1507-1513 (1971).
14. Heidorn, R. F. and H. W. Keller. "Methods for Disposal and Treatment of Plating Room Solutions," *Proc. 13th Purdue Ind. Waste Conf.* **13,** 418-427 (1958).
15. Schore, G. "Electronic Equipment for Use in Automated Water Treatment Systems," Presented at the 27th Ind. Waste Conf., Purdue University (1972).
16. Ross, R. D. (Ed). *Industrial Waste Disposal,* Reinhold Book Corp., (New York: Van Nostrand Reinhold Co., 1968).
17. McGarvey, F. X., R. E. Tenhoor and R. P. Nevers. "Brass and Copper Industry: Exchangers for Metals Concentration from Pickle Rinse Waters," *Ind. Eng. Chem.* **44,** 534-541 (1952).
18. Arden, T. V. *Water Purification by Ion Exchange* (New York: Plenum Publishing Corp., 1968).
19. "Winning Heavy Metals from Waste Streams," *Chem. Eng.* **78,** 62, 64 (April 19, 1971).
20. Walker, C. A., B. F. Dodge and J. F. Madden. "Disposal of Electroplating Wastes by Oneida, Ltd.," *Sewage Ind. Wastes* **26,** 1002-1013 (1954).

21. Sigworth, E. A. and S. B. Smith. "Adsorption of Inorganic Compounds by Activated Carbon," *J. Amer. Water Works Assoc.* **64**, 386-391 (1972).
22. Eastman Kodak Co. "Disposal of Photographic-Processing Waste," Kodak Publication No. J-28, (Rochester, New York: Eastman Kodak Co., 1969).
23. Silman, H. "Treatment of Rinse Water from Electrochemical Processes," *Metal Finishing* **69**, 62-66 (1971).

21

TREATMENT TECHNOLOGY FOR REDUCTION OF TOTAL DISSOLVED SOLIDS

Removal of total dissolved solids (TDS) from wastewaters is one of the more uncommon and, in many cases, most expensive waste treatment procedures. Where high total dissolved solids result from heavy metal or hardness ions, reduction may be accomplished by chemical precipitation methods, as discussed in other chapters. Where the dissolved solids are present, as for an example sodium, magnesium or sulfate ions, total dissolved solids reduction requires more specialized treatment methods, as discussed in this chapter.

INDUSTRIAL SOURCES

Although dissolved solids are present to some extent in any water, natural or waste, most industrial processes and many industrial waste treatment processes, result in increased quantities of TDS. One major exception is where lime is added for anion precipitation. In general, TDS of natural waters consist mainly of carbonates, bicarbonates, chlorides, sulfates, phosphates and nitrates of calcium, magnesium, sodium and potassium, with traces of iron, manganese and other substances.

The mineral content of natural waters is normally raised by the addition of chemical wastes, dissolved salts, acids, and alkalis. For example, TDS levels of metal plating effluents, after waste treatment for acid neutralization and reduction of cyanide and heavy metals, are presented in Table 92. A recent study indicates that TDS in inorganic chemical industry effluents may range from 1000–150,000 mg/l (2). An effluent TDS of 1163 mg/l has been reported from a petroleum industry refinery (3), and TDS in a carpet mill waste of 5850 mg/l, with a raw water supply TDS of 410 mg/l (4). Source water TDS is quite variable, depending upon geographical location and surface versus ground water source, and thus can influence wastewater TDS.

Table 92. Total Dissolved Solids Concentration in Treated Metal Plating Wastes (1).

Plant No.	Location	TDS (mg/l)
1	Illinois	555
2	Michigan	490
3	Illinois	15,100
4	Virginia	906
5	Illinois	1,065
6	California	625
7	Kentucky	810
8	California	435
9	New York	610
10	Indiana	166
11	New York	1,580
12	Arkansas	393
13	Illinois	2,260
14	Ohio	4,066
15	Pennsylvania	793

TREATMENT TECHNOLOGY

The nature of nonprecipitable dissolved solids and the treatment processes available to remove or reduce such solids are such that practically speaking, there are no intermediate levels of treatment. Treatment for removal of dissolved solids, if applied at all, generally produces a water containing very low residual TDS. Partial treatment of a high TDS waste stream thus may involve splitting the stream, treating one portion to remove essentially all dissolved solids, and recombining the treated and untreated streams in the proportion necessary to achieve the desired final solids concentration. Therefore, treatment costs discussed later in this chapter do not refer necessarily to the total waste volume produced by an industrial process, but only to that portion of the waste that must be separated and treated to yield required effluent quality.

The major processes employed for reduction of TDS are reverse osmosis, electrodialysis, distillation, and ion exchange. Ion exchange and electrodialysis are generally applicable for TDS concentrations up to 5000 mg/l, while reverse osmosis and distillation are employed for TDS concentrations to 50,000 mg/l (2). Evaporative ponds may be required at extremely high TDS concentrations, above 50,000 mg/l. Table 93 presents capital and operating costs of TDS treat-

Table 93. Capital and Maintenance and Operating (M&O) Treatment Costs for Reduction of TDS (2).

Flow (MGD)	TDS (mg/l)	Type of Process[a]	Capital Cost of Process ($1000s)	Total[b] Capital Costs ($1000s)	M&O Cost of Process (¢/1000 gal)	Total[b] M&O Cost (¢/1000 gal)
0.5	3,000	(1)	640	1,439	40.0	65.5
		(2)	455	1,240	65.0	97.5
		(3)	806	1,701	98.0	122.5
		(4)	1,700	2,551	25.0	49.5
		(5)	40	40	5.0	5.0
	30,000	(1)	784	1,704	75.0	112.5
		(3)	806	1,713	98.0	130.5
		(5)	40	40	5.0	5.0
	150,000	(5)	40	40	5.0	5.0
1.0	3,000	(1)	1,120	2,185	31.0	51.5
		(2)	896	1,929	54.0	79.3
		(3)	1,610	2,810	85.0	103.0
		(4)	3,020	4,700	33.3	48.3
		(5)	100			5.0
	30,000	(1)	1,340	2,550	50.0	77.6
		(3)	1,610	2,892	85.0	110.3
10.0	3,000	(1)	6,700	10,651	18.5	26.0
		(2)	6,400	10,496	38.0	49.0
		(3)	9,070	13,724	38.0	44.0
		(4)	14,800	21,160	30.0	34.0
		(5)	410	410	0.5	0.5
	30,000	(1)	8,400	13,567	25.0	38.0
		(3)	9,070	14,100	38.0	49.0
50.0	3,000	(1)	29,100	42,356	15.0	19.0
		(3)	34,700	49,920	22.0	26.0

[a] Process: 1) Reverse osmosis; 2) Electrodialysis; 3) Distillation; 4) Ion exchange; 5) Evaporative ponds.

[b] Total costs include prefiltration for removal of suspended solids, and brine waste disposal.

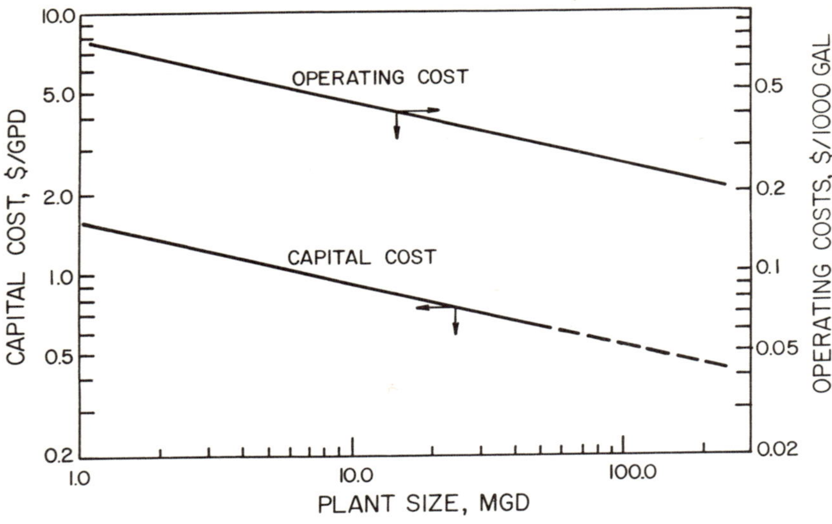

Figure 44. Capital and operating costs for multiple-effect evaporation/distillation (2).

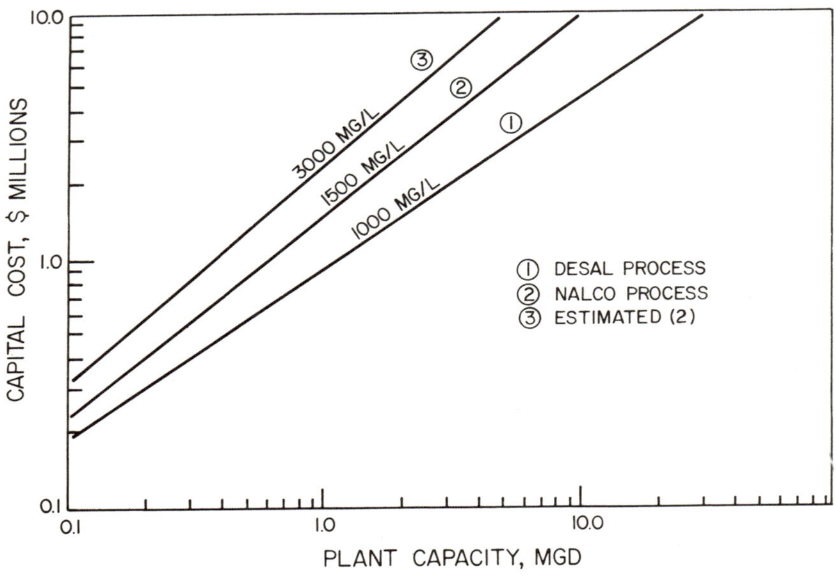

Figure 45. Capital cost of dissolved solids reduction by ion exchange treatment (2).

ment processes, in terms of waste stream volume, and TDS concentrations. Capital and operating costs are also presented in Figures 44-49. Additional cost curves are presented in Figures 50 and 51. The lower three curves in Figure 51 refer to various ion exchange media.

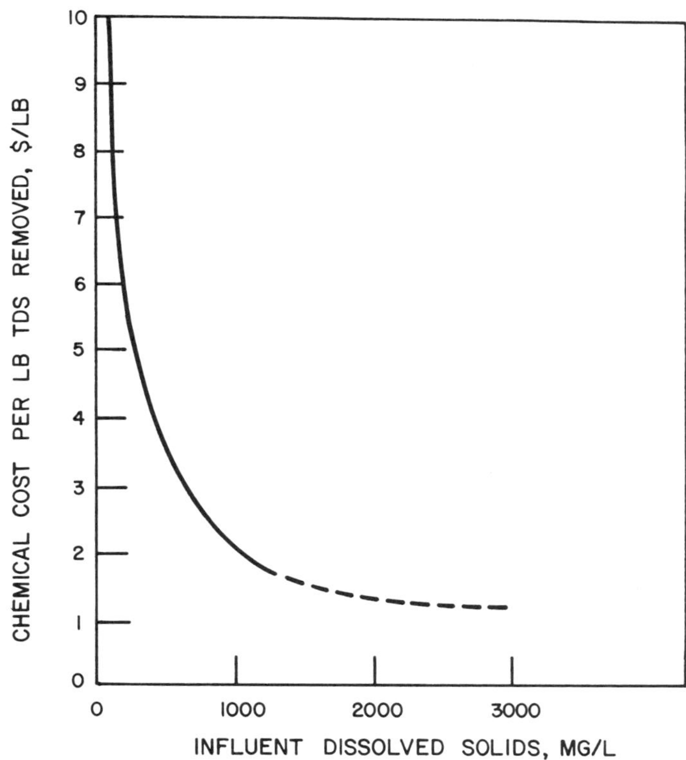

Figure 46. Ion exchange resin regenerating chemicals cost per pound of total dissolved solids removed (2).

All treatment processes for TDS yield a brine solution containing the dissolved solids removed in the process in addition to the treated product water (effluent). Disposal of this brine is usually by deep-well injection or to a solar evaporating lagoon. An important exception exists for those industrial wastes where recovery of a valuable constituent can be combined with waste treatment.

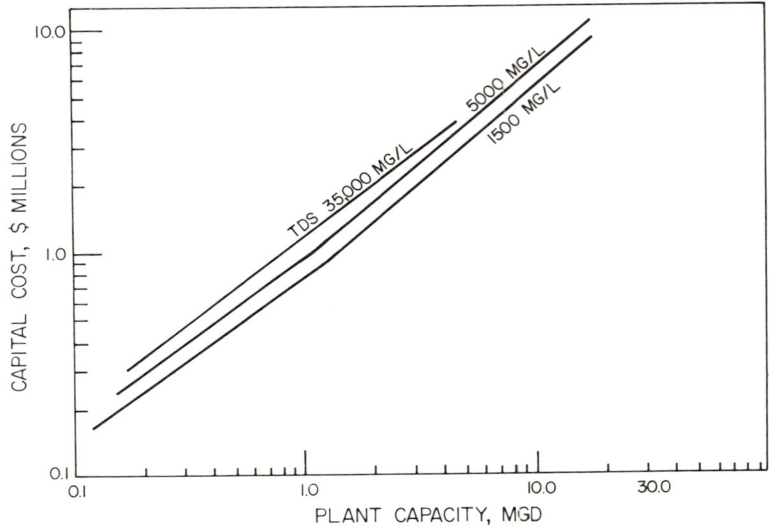

Figure 47. Capital costs for reverse osmosis treatment plants (2).

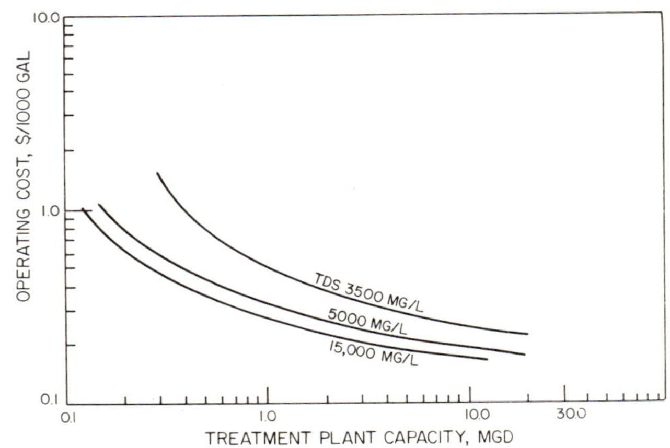

Figure 48. Operating costs for dissolved solids reduction by reverse osmosis (2).

Distillation

Recondensed water vapor from distillation processes contains little inorganic dissolved solids, but may contain volatile organic fractions driven off with the water vapor, and recondensed. Even at 1962 prices (for fuel), water treatment by evaporation averaged

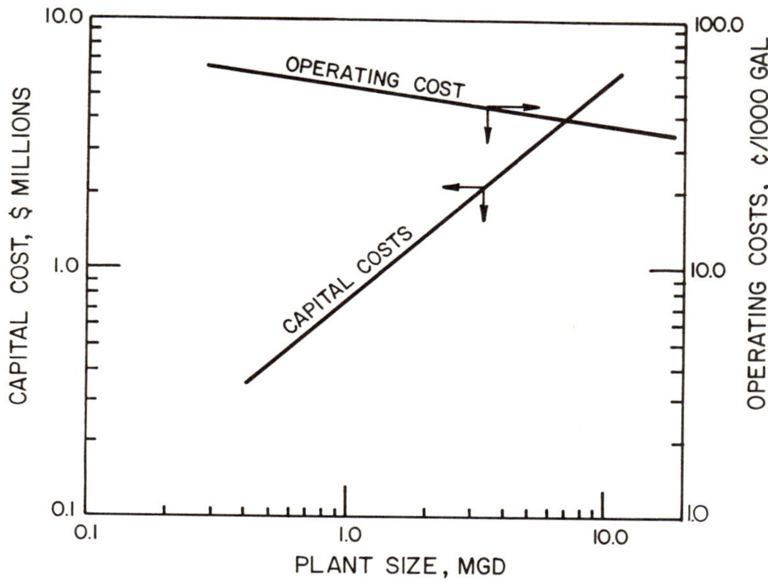

Figure 49. Capital and operating costs for an electrodialysis treatment plant, at an influent total dissolved solids concentration of 3000 mg/l (2).

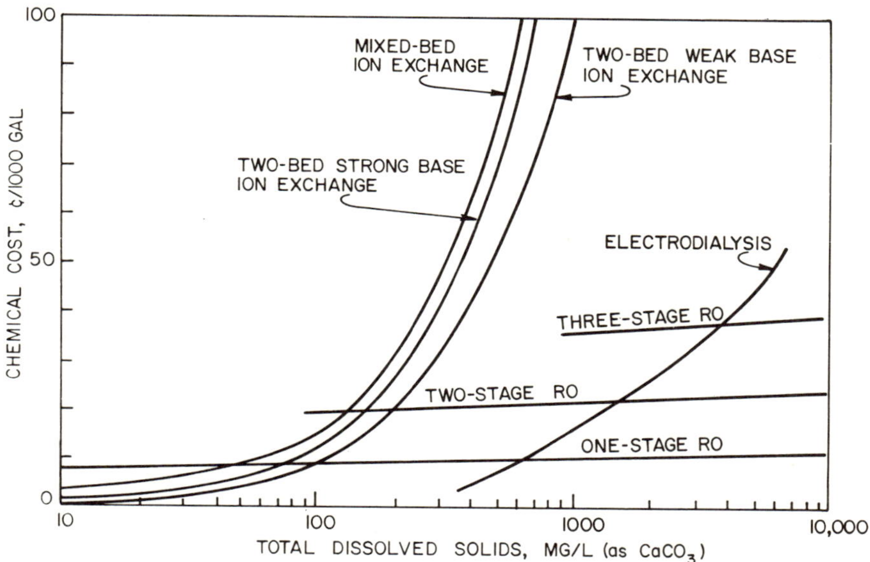

Figure 50. Chemical treatment cost versus total dissolved solids for various demineralization systems. "RO" is reverse osmosis. [From Ahlgren (5), courtesy of *Industrial Water Engineering.*]

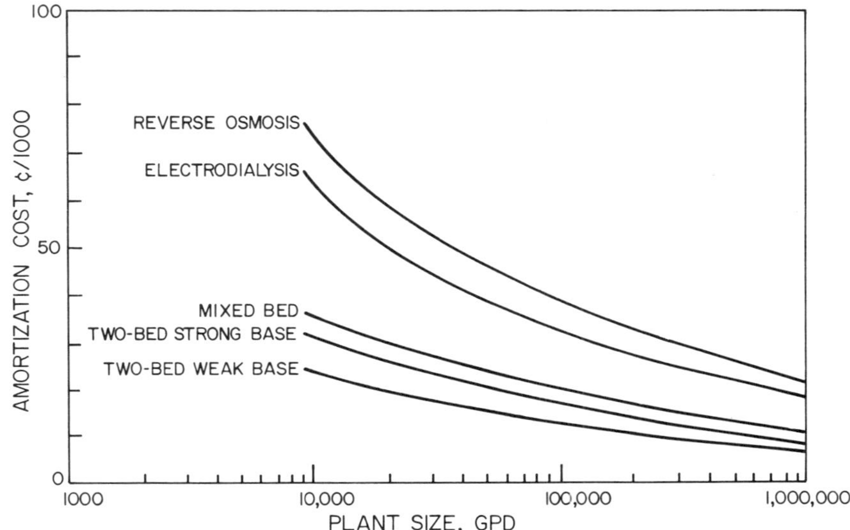

Figure 51. Amortization costs versus plant size for various demineralization systems. Costs are based on 7-yr write-off, 7% interest rate, and 330 days/yr plant operation. [From Ahlgren (5), courtesy of *Industrial Water Engineering.*]

about 50¢/1000 gal (6). Costs shown in Table 93 confirm this estimate, for water of high TDS. At current high fuel prices, distillation is increasingly falling out of favor as a treatment process. Most industrial experience with distillation is to purify raw water for boiler operations.

Prescott (7) has described a 50,000–gpd evaporative process used to treat cooling tower blowdown near El Paso, Texas. The process requires only 20 Btu/lb of water evaporated. The blowdown has a TDS of 2331 mg/l, and the condensed vapors, which are recycled to the cooling towers, contain 8 mg/l TDS. Scaling in the evaporator is prevented by using a seed slurry of brine crystals. The concentrated waste brine from the process is disposed to an evaporation pond. Capital investment for the unit was $400,000, and operating costs are reported at $2.00/1000 gal. This does not include savings in water reuse.

Ion Exchange

Ion exchange resins, depending upon their chemical nature, show preferential selectivity for specific ions. For example, weak-base anion resins are more selective for removing chloride than are strong-base resins. Table 94 presents data on ion selectivity for various types of exchange resins. The greater the affinity of the resin for the adsorbed ion, the more complete removal is achieved. However, a high affinity also means greater difficulty in regenerating the resin. Because regeneration of an effective resin is difficult, it is never carried to completion. As a result, the operational capacity of an ion exchange resin may be only 50–60% of theoretical capacity. For treatment of concentrated wastes, large resin regenerant volumes result, which may present a considerable disposal problem.

Ion exchange processes are very sensitive to both clogging and fouling. An ion exchange resin bed is a good filter and suspended solids in the wastewater will rapidly clog the bed. Fouling results when the resin adsorbs materials which, either because of their affinity or adsorption into the resin pores, cannot be removed in the regeneration step. The Mn^{+4} ion, for example, has such high affinity for certain exchangers that it is almost impossible to remove by conventional regeneration. The net result is loss of resin reactivity.

Ion exchange has been widely successful for water-softening and boiler-water deionization. It has been less successful, however, in industrial wastewater treatment, with many reports of irreversible fouling, poor removal efficiency and similar difficulties. Its most successful industrial wastewater use to date has been with metal plating rinse waters.

Table 94. Ion Exchange Resins Selectivity (8).

Resin	Resin Selectivity[a]
Strong-acid cation (Sulfonic)	Li^+, H^+, Na^+, NH_4^+, K^+, Rb^+, Cs^+, Mg^{+2}, Zn^{+2}, Cu^{+2}, Ca^{+2}, Pb^{+2}
Weak-acid cation (Carboxylic)	Na^+, K^+, Mg^{+2}, Ca^{+2}, Cu^{+2}, H^+
Strong-base anion (Type I)	F^-, OH^-, $H_2PO_4^-$, HCO_3^-, Cl^-, NO_2^-, HSO_3^-, CN^-, Br^-, NO_3^-, HSO_4^-, I^-
Strong-base anion (Type II)	F^-, $H_2PO_4^-$, OH^-, HCO_3^-, Cl^-, NO_2^-, HSO_3^-, CN^-, Br^-, NO_3^-, HSO_4^-, I^-
Weak-base anion	F^-, Cl^-, Br^-, I^-, PO_4^{-3}, NO_3^-, CrO_4^{-2}, SO_4^{-2}, OH^-

[a] Increasing selectivity left to right.

Ion exchange has been used to purify a brackish municipal water supply, which contained 1000 mg/l of sulfate and several hundreds of mg/l of carbonates (9). The effluent water contains less than 250 mg/l of dissolved solids. Treatment costs are 20¢/1000 gal with a capital investment of $700,000 for the 0.5-MGD plant. This is much less than the capital cost indicated in Table 93 for an equivalent waste. The two-stage ion exchange process is regenerated by lime and sulfuric acid. Large amounts of calcium sulfate sludge (gypsum) are produced, and disposed of by landfill (9).

Higgins (10) has presented costs of ion exchange purification of a secondary sewage treatment plant effluent totaling 10 MGD. Capital costs were estimated at $1,963,000 and amortization cost at 4.9¢/1000 gal. Operating costs were 6.87¢/1000 gal or $68.70 per day. This plant would effectively treat a waste containing dissolved solids of 683 mg/l. Young has reported ion exchange treatment costs of about 9¢/1000 gal, while amortization costs were between 15¢ and 20¢/1000 gal (6). Table 93 indicates much higher operating costs for ion exchange treatment, at 25–50¢/1000 gal.

Reverse Osmosis

The reverse osmosis process produces desalted effluent by forcing water through semipermeable membranes at high pressure. These membranes are more permeable to pure water than to dissolved solids. Dissolved solids removal is not complete, as shown in Figure 52. Effluent TDS content increases with increasing wastewater dissolved solids levels. In addition, low molecular weight organic compounds will pass the membrane as effectively as water. Phenol (MW 94) is a good example. Its concentration may be 10–20% greater in the permeate than feedwater, even for a 97% rejection of NaCl membrane (12). Removals of up to 95% inorganic TDS are obtainable, as long as the membranes are in good condition (13). At acidic wastewater pH or pH above 9, the reverse osmosis membranes will hydrolyze and dissolve in a matter of hours. Higher temperature liquids, above 100°F, may also shorten membrane life (11,14). Other factors which affect the useful life of reverse osmosis membranes are excessive pressure (above 500 psig), bacterial degradation and precipitate formation at the membrane surface. Of these five factors, only precipitation is reversible. The adverse effect of pH and temperature, in particular, prevents the use of reverse osmosis on many industrial wastewaters. For example, most effec-

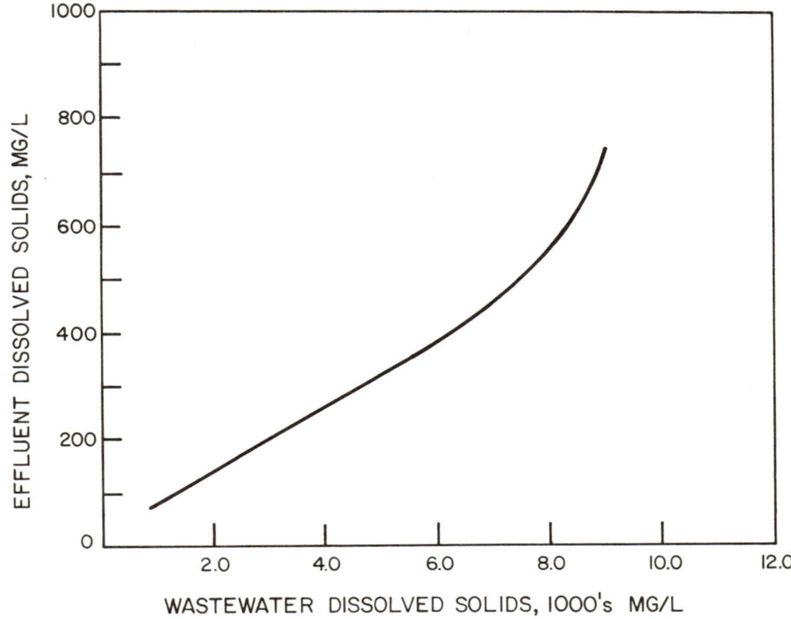

Figure 52. Relationship of effluent to influent total dissolved solids concentration for reverse osmosis treatment (11).

tive cyanide removal by reverse osmosis occurs on pH above 10.0. However at this pH, alkaline hydrolysis rapidly destroys the membrane.

Good results have been obtained in purifying brackish water containing up to 20,000 mg/l dissolved solids, with operating costs estimated at 36–19¢/1000 gal for 1- to 5-MGD plants, respectively (15). Detailed operating results are presented in Table 95 for three full-scale reverse osmosis plants. Plants A and B, at 24,000 and 150,000 gpd, treated brackish water for domestic consumption. Plant C treated municipal sewage treatment plant effluent at 10,000 gpd. The ranges of percent removals for various ions monitored are also contained in Table 95. Although efficiency of removal was not consistent for all minerals, TDS reductions exceeding 90% were found. Table 96 contains additional cost information on reverse osmosis treatment. These values, which are more recent than those shown in Figure 47 and 48, indicate lower capital costs but much greater

Table 95. Operating Results for Full-Scale Reverse Osmosis Plants.

	Plant A (16)		Plant B (16)		Plant C (17)		Range of Percent
Mineral (mg/l)	Influent	Effluent	Influent	Effluent	Influent	Effluent	Removal
Calcium	312	10	160	2	68	0.8	96.8–99.8
Magnesium	141	12	49	4	12	1.5	87.5–91.8
Sodium + Potassium	60	15	410	41	139	15.3	75.0–90.0
Bicarbonate	183	37	297	59	41	22	46.4–79.8
Chloride	96	59	351	32	260	15	29.5–94.2
Sulfate	1008	21	864	29	85	Nil	96.6–100.0
Iron	.05	Nil	2.9	.05	.02	Nil	98.3–100.0
TDS	1008	155	2100	170	756	24	91.4–96.8

Table 96. Representative Costs of Reverse Osmosis Treatment (18).

Cost Component	Treatment Cost (¢/1000 gal)		
	1-MGD	10-MGD	100-MGD
Capital recovery (6%, 25 yr)	10	7.8	5.8
Power (1¢/kwh)	6.4	6.4	6.4
Chemicals	2.4	2.4	2.4
Operation and maintenance	5.4	1.8	0.6
Membrane replacement			
2-yr life	21.3	17.0	17.0
3-yr life	13.8	11.0	11.0
Total operating cost			
2-yr life	45.5	35.4	32.2
3-yr life	38.0	29.4	26.2
Capital cost (million $)	0.47	3.66	27.5

operating costs. Recent (1971) estimates place the operating costs of reverse osmosis at 30–60¢/1000 gal (13).

Electrodialysis

The amount of dissolved solids removed by electrodialysis depends upon the retention time of the waste in the electrodialysis unit among other engineering factors. Thus, the degree of purification achieved by this process, unlike those described previously, can be selected to a degree. Electrodialysis uses ion exchangers in membrane form, with an electrical driving force to promote ion transport across the membrane. Ions are thus removed from solution, leaving water behind. By contrast to ion exchange, which must undergo periodic regeneration, electrodialysis is essentially a continuous process. Since electric current must pass through the liquid, high resistance is encountered as the water becomes increasingly pure.

The operation of an electrodialysis plant in Buckeye, Arizona, has been described (19). This plant, which has been in operation since 1962, reduces TDS of approximately 2100 mg/l to about 500 mg/l. The plant is used to produce municipal water for the town of Buckeye, from brackish ground water which would otherwise be too high in TDS for consumption. Operating costs for this electrodialysis plant of 4.8¢/1000 gal of treated water produced have been reported (20). The 1964 cost of a 10–20-MGD electrodialysis plant has been estimated at 9–10¢/1000 gal (21). This included amortization plus operating and maintenance. These costs contrasted with 1962 cost data of up to 20¢/1000 gal.

SUMMARY

Most TDS treatment processes, with the exception of industrial waste precipitation, have been developed either to purify saline or brackish waters or to provide deionized boiler feed and manufacturing process waters. There is little information in the technical literature on industrial wastewater TDS removal by processes discussed in this chapter. The inorganic chemicals industries appear to be the major users of nonprecipitation methods such as ion exchange and reverse osmosis to reduce TDS in wastewaters. Costs of TDS removal are high compared to many other types of industrial waste treatment processes. However, often only a fraction of the waste stream need be treated to achieve intermediate TDS levels in the waste.

REFERENCES

1. Zievers, J. F., R. W. Crain and F. G. Barclay. "Waste Treatment Metal Finishing: U.S. and European Practices," *Plating* **55**, 1171-1179 (1968).
2. *The Economics of Clean Water, Vol. III. Inorganic Chemicals Industry Profile,* (Washington, D.C.: U.S. Dept. of the Interior, 1970).
3. Benger, M. "The Disposal of Liquid and Solid Effluents from Oil Refineries," *Proc. 21st Purdue Ind. Waste Conf.* **21**, 759-767 (1966).
4. Stone, R. "Walter Carpet Mill Industrial Waste System," *J. Water Poll. Cont. Fed.* **44**, 470-478 (1972).
5. Ahlgren, R. M. "Membrane vs. Resinous Ion Exchange Demineralization," *Ind. Water Eng.* **8**, 12-14 (1971).
6. Young, L. "Ion Exchange Softening Process Operation and Maintenance," *Proc. 25th Annual Short Course of Superintendents and Operators of Water, Sewage and Industrial Waste Disposal Systems,* 103-106 (1962).
7. Prescott, J. H. "New Evaporation-Step Entry," *Chem. Eng.* **78** (29), 30-31 (1971).
8. Calmon, C. "Modern Ion Exchange Technology," *Ind. Water Eng.* **9** (3), 12-15 (1972).
9. "Ion Exchangers Sweeten Acid Water," *Environ. Sci. Technol.* **5**, 24-25 (1971).
10. Higgins, I. R. "A Unique Process for Demineralizing Waste Water," *Ind. Water Eng.* **2**, 26-29 (1965).
11. Morris, D. C., W. R. Nelson and G. O. Walraven. "Recycle of Papermill Waste Waters and Applications of Reverse Osmosis," U.S. EPA Report 12040 FUB 01/72 (1972).
12. Spatz, D. D. "Electroplating Waste Water Processing with Reverse Osmosis," *Prod. Finishing* **36** (11), 79-89 (1972).
13. Culp, R. L. and G. L. Culp. *Advanced Wastewater Treatment,* (New York: Van Nostrand Reinhold Co., 1971).

14. "Reverse Osmosis Module Operates Below 450 PSI," *Chem Eng.* **78** (9) 72-73 (1971).
15. Channabasappa, K. C. and F. L. Harris. "Economics of Large-Scale Reverse Osmosis Plants," *Ind. Water Eng.* **7,** 20-24 (1970).
16. Shields, C. P. "Reverse Osmosis for Municipal Water Supply," *Water Sewage Works* **119,** 64-70 (1972).
17. Nusbaum, I., J. H. Sleigh, Jr. and S. S. Kremen. "Study and Experiments in Waste Water Reclamation by Reverse Osmosis," U.S. EPA Report 17040—05/70 (1970).
18. Cruver, J. E. and I. Nusbaum. "Application of Reverse Osmosis to Wastewater Treatment," Presented at 27th Ind. Waste Conf., Purdue University (1972).
19. Oliver, J. C. "The Port Mansfield Desalting Plant," *Public Works* (June, 1965).
20. Gilliland, E. R. "The Current Economics of Electrodialysis," Presented at 1st Internat. Symp. of Water Desalination, Washington, D.C. (October, 1965).
21. Smith, J. D. and J. L. Eisenmann. "Electrodialysis in Waste Water Recycle," *Proc. 19th Purdue Ind. Waste Conf.* **19,** 738-760 (1971).

22

TREATMENT TECHNOLOGY FOR ZINC

INDUSTRIAL SOURCES

Industries discharging waste streams which carry significant quantities of zinc include:

- Steel works with galvanizing lines
- Zinc and brass metal works
- Zinc and brass plating
- Silver and stainless steel tableware manufacturing
- Viscose rayon yarn and fiber production
- Groundwood pulp production
- Newsprint paper production

Zinc salts are also used in the inorganic pigments industry, and high zinc levels have been reported in acid mine drainage water.

The primary source of zinc in wastewaters from plating and metal processing industries is the drag-out solution adhering to the metal product after removal from pickling or plating baths. The metal is washed free of this solution, and the contaminants are transferred to the rinse water. The pickling process consists of immersing the metal (zinc or brass) in a strong acid bath to remove oxides from the metal surface. Finished metals are brightened by submergence in a bright dip bath containing strong chromate concentrations as well as acid.

Plating solutions typically contain 5,000–34,000 mg/l of zinc. These concentrated solutions may be discharged periodically, due to contamination. The concentration of zinc in a plating rinse water will be a function of the bath zinc concentration, drainage time over the bath, and the volume of rinse water used. Zinc and brass plating and rinse solutions generally also contain cyanide. Waste concentrations of zinc range from less than 1 to more than 1000 mg/l in various waste streams described in the literature. However, average values seem to fall between 10 and 100 mg/l. Table 97 summarizes values reported for various zinc-bearing wastewaters.

TREATMENT TECHNOLOGY

Treatment processes employed for wastewater zinc removal may involve either chemical precipitation, with disposal of the resultant sludge, or recovery. Recovery processes include ion exchange and evaporative recovery, but may also be precipitation processes, where a relatively pure zinc sludge is reclaimed. Recovery of plating wastes frequently proves to be more economical on an overall basis than conventional precipitation and sludge disposal.

Chemical Precipitation

The precipitation process most frequently involves adjustment of pH to achieve alkaline conditions, and precipitation of zinc hydroxide with either lime or caustic. Lime addition has been the most widely accepted method for pH adjustment, in spite of the concurrent precipitation of calcium sulfate in the presence of high sulfate levels in pickling bath wastes. The precipitation of calcium sulfate along with the zinc hydroxide increases the total amount of sludge to be disposed of.

Table 98 is a representative summary of published treatment results for hydroxide precipitation. These values reflect a wide range of industrial systems and generally, treatment is not for zinc alone. Where cyanide or chromate is also present in the waste, as frequently occurs in zinc and brass plating, cyanide destruction and chromate reduction must precede metal hydroxide precipitation. Settling efficiency affects effluent concentrations, as is evident by improvements in effluent zinc levels resulting from sand filtration of settled effluents. Incomplete cyanide treatment will increase effluent zinc levels, due to complexation (21).

The use of hydrogen sulfide at pH 2 has been reported to achieve "complete removal" of zinc from an electrolytic nickel recovery system (22). Sodium sulfide treatment of a gold ore refining waste, initially containing zinc at 42 mg/l, achieved precipitation to 1.16 mg/l at a treatment pH of 7.0 (23). However, treatment of the

Table 97. Concentrations of Zinc in Process Wastewaters.

Industrial Process	Zinc Concentration (mg/l)		
	Range	Average	Reference
Metal Processing			
Bright dip wastes	0.2-37.0		1
Bright mill wastes	40-1,463		2
Brass mill wastes	8-10		2
Pickle bath	4.3-41.4		1
Pickle bath	0.5-37		2
Pickle bath	20-35		3
Aqua fortis and CN dip	10-15		3
Wire mill,pickle	36-374		4
Plating			
General	2.4-13.8	8.2	5
General	55-120		6
General	15-20	15	7
General	5-10		3
Zinc	20-30		3
Zinc	70-150		1
Zinc	70-350		8
Brass	11-55		1
Brass	10-60		8
General	7.0-215	46.3	1
Plating on zinc castings	3-8		3
Galvanizing of cold rolled steel	2-88		9
Silver Plating			
Silver bearing wastes	0-25	9	1
Acid waste	5-220	65	1
Alkaline	0.5-5.1	2.2	1
Rayon Wastes			
General	250-1000		10
General	20[a]		11
General	20-120		12
Other			
Vulcanized fiber	100-300		13
Cooling tower blowdown	6		14

[a]After process recovery of zinc by ion exchange.

Table 98. Summary of Hydroxide Precipitation Treatment Results for Zinc Wastewaters.

Industrial Source	Zinc Concentration (mg/l) Initial	Final	Comments[a]	Reference
Zinc plating	—	0.2-0.5		15
General plating	18.4	2		16
General plating		0-6	Sand filtration	17
General plating	55-120	⩽1.0 (ave)		6
Vulcanized fiber	100-300	⩽1.0		13
Brass wire mill	36-374	0.08-1.60	Integrated treatment for copper recovery	4
Tableware plant	16.1	0.02-0.23	Sand filtration	1
Viscose rayon	20-120	0.88-1.5		12
Viscose rayon	70	3-5	Caustic	18
Viscose rayon	20	1.0		11
Metal fabrication	—	0.5-1.2	1) Sedimentation	19
	—	0.1-0.5	2) Sand filtration	
Blast furnace gas scrubber water	50	0.2		20

[a] All treatment involved precipitation plus sedimentation. Special or additional aspects of treatment are indicated under "Comments."

same waste by caustic at pH 10 reduced zinc to 0.02 mg/l. Treatment of acid mine drainage containing zinc at 122–294 mg/l by first-stage lime addition to pH 5.0–5.5, followed by second-stage sulfide addition to pH 5.5–6.4 produced variable effluent zinc levels ranging from 0.2–30.0 mg/l (24). These results suggest that sulfide precipitation cannot achieve equivalent levels of treatment to hydroxide precipitation.

The patented integrated treatment system has been claimed to achieve effluent zinc levels below 1.0 mg/l in plating and pickling systems (25). The process involves immersion of a freshly plated part in a concentrated soda ash solution, prior to rinsing of the part. In one reported application of the process in a small plating shop, effluent rinse water zinc was reduced from 2.0–5.0 mg/l to 0.29–0.44 mg/l after installation of the process (26). The treatment bath contained hydrazine, soda ash and caustic, and was maintained at pH 9.5–10.5. Hydrazine was present to reduce trace amounts of cupric ion and allow precipitation of cuprous oxide.

A process has been developed by DuPont primarily for small zinc and cadmium-cyanide plating operations (27). A proprietary formulation called Kastone (composed of a catalyst, hydrogen peroxide and stabilizers) oxidizes the cyanide and promotes the formation of metal oxide instead of hydroxide. The former are readily removed by filtration apparatus, which most small plating plants already have in operation. Effluent treatment levels of 0.29–2.9 mg/l zinc have been reported for zinc plating rinse waters (28).

Costs for zinc removal systems cannot usually be isolated from the cost of treatment for other constituents contained in the particular waste. Neutralization must be accomplished even in waste streams containing no zinc; cyanide must be oxidized and chromium reduced. Costs directly related to zinc treatment result from the lime requirements, clarifier costs in the absence of other metal hydroxides, and dewatering and disposal costs per ton of hydroxide sludge produced. In the past, precipitation of metal hydroxides was relatively inexpensive, but costs have increased considerably in achieving more effective treatment, generally due to increased costs of better clarifiers or polishing of effluents with sand filtration (29).

Ross (30) estimated operating costs for lime treatment of 35,000 gpd of a zinc-nickel waste, containing 5.5 mg/l of nickel and 3.5 mg/l of zinc, to be $0.34/1000 gal. Lime quantity was estimated at 185 lb per day and sludge disposal amounted to approximately $0.03/1000 gal of wastewater treated. The range of costs associated with various precipitation based treatment systems is given in Table 99. Unless otherwise specified, amortization costs are not incorpo-

Table 99. Costs for Precipitation Processes.

Wastewater Volume (1000 gpd)	Treatment Process	Capital Costs ($/1000 gpd)	M&O Costs (¢/1000 gal)	Ref.
720	Batch-flow separation	789	16.6	17
720	Batch-no flow separation	643	19.4	17
720	Continuous-flow separation	874	15.3	17
720	Continuous-no flow separation	564	18.6	17
400	Continuous-zinc recovery, 50,000 lb/mo	3000	na	13
240-360 (gpd)	Continuous (exclusive of collection system)	300-900	7-25	9
1728	Continuous-zinc recovery (includes collection system)	422	0 based upon 2000 lb/day Zn recovery	12
216	Continuous	2768	179	4
	Integrated-recovered water reuse	658	72	4
30 (gpd)	Continuous	9200	379	6
na	Zinc-Nickel Continuous-lime	na	34	30

rated in these figures. Some of these plants were designed for direct zinc recovery, others were designed for product recovery other than zinc. A major cost of treatment facility installation can be waste collection and transfer systems. A range of 21–45% of total plant investment has been attributed to the collection system (17). Flow separation collection systems may be more than twice as costly as combined systems.

Some general capital costs for plating waste disposal plants are presented in Figure 53, as a function of plant capacity. Although somewhat out of date, general economy of scale relationships are illustrated. Simple precipitation costs should be around $0.25/1000 gal without filtration and $0.45/1000 gal with filtration (32). Additional treatment will increase costs accordingly. An example is

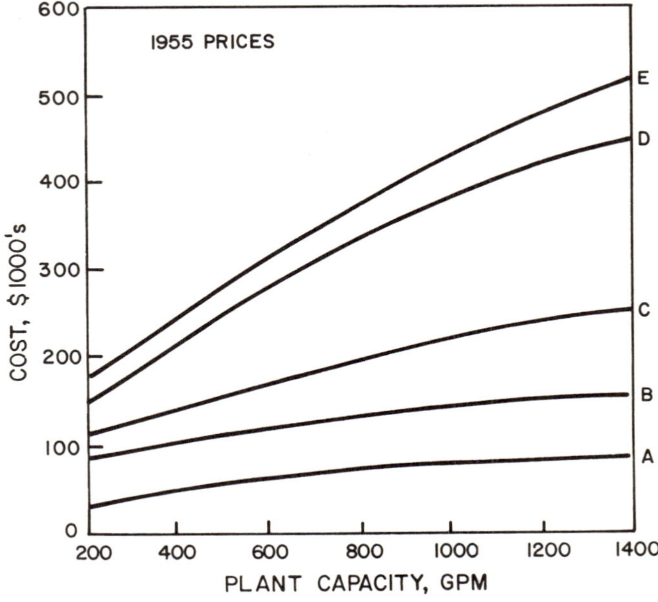

Figure 53. Approximate cost of treatment plant for plating wastes. Curve A—mixing, flocculation and sedimentation equipment only, installed. Curve B—Curve A plus chemical feeders, pumps and automatic control equipment, installed. Curve C—Curve B plus piping. Curve D—Curve C plus buildings for control, chemical storage, etc. Curve E—Curve D plus engineering, total plant cost. [From Graham (31), courtesy of Van Nostrand Reinhold Co.]

chromium removal, which can range from \$0.45–\$1.90/1000 gal. The data in Table 99 reflect these generalizations and the economy of scale, both in capital and operating costs.

Ion Exchange

Although considered a recovery process, ion exchange may be used as a polishing process for final effluent; or as a means for water reuse, while providing a more concentrated waste for effective treatment. Unless associated with other materials, which are costly to treat and replace, zinc recovery for the metal alone is usually not economical, except at high zinc concentrations. Under conditions of high levels of relative pure zinc, precipitation and acid regeneration of the zinc hydroxide is satisfactory. This situation has been reported in viscous rayon manufacturing (33). Zinc recovery by ion exchange in viscous rayon manufacturing has also been used (10,11), with the acid regenerant reused directly without purification. High residual effluent zinc concentrations have been experienced. In one case (11), residual zinc of 20 mg/l in the ion exchange effluent necessitated further treatment.

Laboratory studies have indicated that at flow rates of 2.5 gal/hr/ft^2, ion exchange was not as effective as lime precipitation for zinc removal (34). A published cost comparison of the treatment of zinc and nickel wastes by ion exchange versus lime treatment indicated ion exchange to be more than twice as costly as lime treatment for a waste flow of 35,000 gpd. Precipitation costs were \$0.34/1000 gal. Recovery value from the ion exchange process was credited for water reuse only (30). Pilot plant studies have been reported to reduce zinc in cooling tower blowdown from 6 to 0.4 mg/l (35).

Evaporative Methods and Process Revision

The major source of zinc-bearing wastewaters is from metal processing and plating operations and specifically from pickling or plating drag-out into rinse water, although periodic dumping of the concentrated plating bath solution may occur as a result of contaminant buildup. A major problem in treating dilute wastewater is the large volumes involved, and attendant capital and operation costs. Direct recycle or evaporative recycle of pickling and plating bath rinse water is made feasible by countercurrent rinse flow.

Rinse water volumes can be reduced to ⅓ to ¼ of normal volumes by the use of cascade or countercurrent rinse practices (25). This reduction in volume yields higher zinc concentrations in the rinse water. One plant was able to reduce zinc cyanide plating rinse requirements from 310 to 12 gph for two units by employing countercurrent rinse (5). The final rinse waters were of sufficient zinc concentration to allow evaporative concentration and recovery of the plating drag-out. Final rinse drag-out required no further treatment, due to the low zinc levels present. Complete elimination of drag-out by evaporative recovery results in impurity build-up in the plating baths, and subsequent need for periodic dumping of the bath solutions.

Culotta and Swanton (29) describe treatment of a zinc and brass plating wastewater by evaporative recovery. Plant modifications reduced process water from 3000 gph to 50 gph and resulted in process chemical savings of $18,000/yr, plus equal savings in chlorine and other treatment chemicals. Capital and operating costs associated with such a closed-loop system were reported to be approximately the same as for destructive recovery.

Application of evaporative recovery has been reported for a plating plant (36). A 300 gph double-effect evaporator was installed on the zinc cyanide plating line. Cost analysis based on 4000 hr/yr operation yielded operating costs of $10,800/yr. For a 10 gph plating bath drag-out, savings of $18,000/yr in chemical costs were claimed. The annual net savings over operation was $7200.

Capital costs for such a system can be approximated from the data of Culotta and Swanton (14) who quote $6000, $7500, and $8000/1000 gpd respectively for single-effect, double-effect and vapor recompression evaporators installed. For a 300-gph double-effect system treatment costs would be $6300/yr, exclusive of savings related to process water reduction. These were estimated to offset treatment costs by $1000/yr (37).

SUMMARY

Zinc removal costs often cannot be isolated from overall waste treatment costs for industrial waste streams, which frequently involve mixtures of other heavy metals and chemicals also requiring treatment. Where zinc removal is the only consideration and recovery is not warranted, removal by precipitation can be accomplished by standard pH adjustment, lime addition, precipitation and flocculation, and sedimentation, employing standard waste treatment equipment. Operational data for existing chemical precipitation units indicate that levels below 1 mg/l of zinc are readily obtainable with lime precipitation, although assurance of consistent removal of precipitated zinc from the effluent stream may require sand filtration and adequate pH control. Approximate costs for general sand filtration treatment are given in Figure 54.

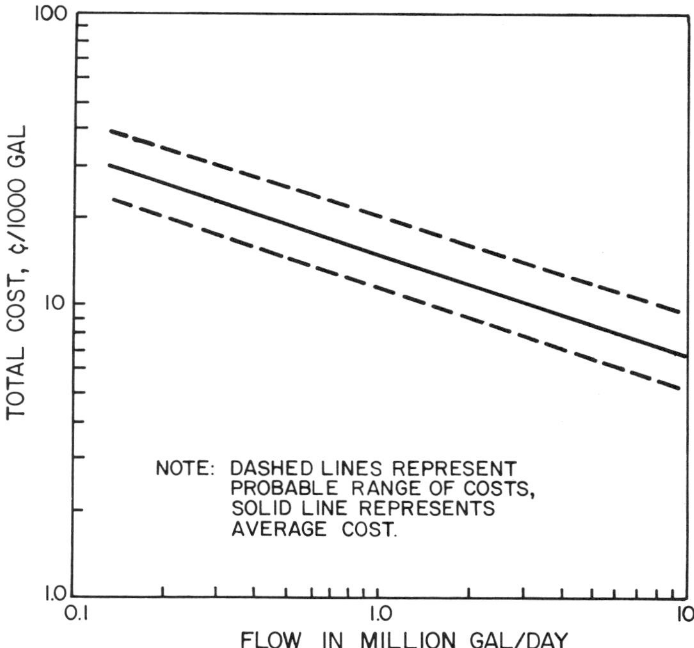

Figure 54. Total treatment costs for coagulation, sedimentation, and rapid sand filtration. Cost includes capital investment (4%, 30 yr), labor, power, chemicals, maintenance and repairs, and heating of building (38).

Other treatment alternatives exist and may be more economical. Among these are process modification for evaporative and ion exchange recovery, although existing process layout and current treatment facilities may cause costs to vary widely with industrial sites and nature of product. Evaporative recovery with countercurrent rinsing to reduce waste volumes, has proved successful on both an economic and zinc removal efficiency basis.

REFERENCES

1. Nemerow, N. L. *Theories and Practices of Industrial Waste Treatment* (Reading, Massachusets: Addison-Wesley Publishing Co., 1963).
2. McGarvey, F. X., R. E. Tenhoor and R. P. Nevers. "Brass and Copper Industry: Cation Exchangers for Metals Concentration from Pickle Rinse Waters," *Ind. Eng. Chem.* **44,** 534-541 (1952).
3. Lowe, W. "The Origin and Characteristics of Toxic Wastes with Particular Reference to the Metal Industries," *Water Poll. Cont. (London)* 270-280 (1970).
4. Volco Brass and Copper Co. *Brass Wire Mill Process Changes and Waste Abatement, Recovery and Reuse,* Water Pollution Control Research Series, #12010 DFP, (Washington, D.C.: U.S. Environmental Protection Agency, 1971).
5. Barnes, G. E. "Disposal and Recovery of Electroplating Wastes," *J. Water Poll. Cont. Fed.* **40,** 1459-1470 (1968).
6. Nyquist, O. W. and H. R. Carroll. "Design and Treatment of Metal Processing Wastewaters," *Sewage Ind. Wastes* **31,** 941-948 (1959).
7. Pinkerton, H. L. "Waste Disposal," In *Electroplating Engineering Handbook,* 2nd ed., A. Kenneth Graham, Ed.-in-chief (New York: Van Nostrand Reinhold Co., 1962).
8. *State of the Art: Review on Product Recovery,* Water Pollution Control Research Series (Washington, D.C.: U.S. Dept. of the Interior, 1970).
9. Donovan, E. J., Jr. "Treatment of Wastewater for Steel Cold Finishing Mills," *Water Wastes Eng.* **7,** F22-F25 (1970).
10. McGarvey, F. X. "The Application of Ion Exchange Resins to Metallurgical Waste Problems," *Proc. 17th Purdue Ind. Waste Conf.* 289-304 (1952).
11. Sharda, C. P. and K. Namwannan. "Viscose Rayon Factory Wastes and Their Treatment," *Technol. Sindri* **3,** 58-60 (1966); *Water Poll. Abstr.* **41,** No. 1698 (1968).
12. American Enka Co., *Zinc Precipitation and Recovery from Viscose Rayon Wastewater,* Water Pollution Control Research Series, #12090 ESG (Washington, D.C.: U.S. Environmental Protection Agency, 1971).
13. "Reclaiming Zinc from an Industrial Waste Stream," *Environ. Sci. Technol.* **6,** 880-881 (1972).
14. Culotta, J. M. and W. F. Swanton. "Recovery of Plating Wastes: Selection of Lowest Cost Evaporator," *Plating* **57,** 1121-1123 (1970).
15. "Effluent Treatment at B. S. R.," *Metal Finishing J.* **17,** 248 (1971).
16. Chalmers, R. K. "Trade Effluent Treatment at the Cannock Factory of Joseph Lucas Limited," *J. Proc. Inst. Sewage Purif.* 357-359 (1965).
17. Chalmers, R. K. "Pretreatment of Toxic Wastes," *Water Poll. Cont. (London)* 281-291 (1970).
18. Rock, D. M. "Hydroxide Precipitation and Recovery of Certain Metallic Ions from Waste Waters," Presented at Annual Meeting,

American Institute for Chemical Engineers, Chicago, Illinois (November-December, 1970).

19. Stone, E. H. F. "Treatment of Non-ferrous Metal Process Waste at Kynoch Works, Birmingham, England," *Proc. 25th Purdue Ind. Waste Conf.* 848-865 (1967).
20. Harrison, J. L. "Iron and Steel Works Pollution Control: Water and Effluents," *Steel Times* **202** (9), 557-567 (1974).
21. Minear, R. A. and J. W. Patterson. "Treatment of Metallic Wastewaters," *Proc. of the Symposium on Water Poll. in Metropolitan Areas,* Illinois Institute of Technology, Chicago, Illinois (November 30, 1972).
22. Banerjee, N. G. and T. Banerjee. "Recovery of Nickel and Zinc from Refinery Waste Liquors: Part I—Recovery of Nickel by Electrodeposition," *J. Sci. Ind. Res.* **11B**, 77-78 (1952).
23. Rosehart, R. and J. Lee. "Effective Methods of Arsenic Removal from Gold Mine Wastes," *Can. Mining J.* 53-57 (June, 1972).
24. Larsen, H. P., J. K. P. Shou and L. W. Ross. "Chemical Treatment of Metal-Bearing Mine Drainage," *J. Water Poll. Cont. Fed.* **45** (8), 1682-1695 (1973).
25. Lancy, L. E. "An Economic Study of Metal Finishing Waste Treatment," *Plating* **54,** 157-161 (1967).
26. Martin, J. J., Jr. "Chemical Treatment of Plating Waste for Removal of Heavy Metals," U.S. Environmental Protection Agency, Report EPA-R2-73-044 (May, 1973).
27. "New Process Detoxifies Cyanide Wastes," *Environ. Sci. Technol.* **5,** 496-497 (1971).
28. Lawes, B. C., L. B. Fournier and D. B. Mathre. "A Peroxygen System for Destroying Cyanides in Zinc and Cadmium Electroplating Rinse Waters," *Plating* **60** (9), 902-909 (1973).
29. Culotta, J. W. and W. F. Swanton. "The Role of Evaporation in the Economics of Waste Treatment for Plating Operations," *Plating* **55,** 957-961 (1968).
30. Ross, R. D. *Industrial Waste Disposal,* (New York: Van Nostrand Reinhold Company, 1968).
31. Graham, A. K. (Ed). *Electroplating Engineering Handbook* (New York: Van Nostrand Reinhold Co., 1962).
32. Anderson, J. S., J. M. Phillips and C. B. Schriver. "Water Contamination by Certain Heavy Metals," Presented at the 44th Annual Meeting of the Water Pollution Control Federation, San Francisco, California (October 5, 1971).
33. Saxena, K. L. and R. N. Chakraboryt. "Viscose Rayon Wastes and Recovery of Zinc Therefrom," *Technol. Sindii* **3,** 29-33 (1964); *Water Poll. Abstr.* **41,** No. 1699 (1968).
34. Kantawala, D. and H. D. Tomlinson. "Comparative Study of Recovery of Zinc and Nickel by Ion Exchange Media and Chemical Precipitation," *Water Sewage Works III,* R281-R286 (1964).
35. Chamberlain, D. G. and R. E. Anderson. "Selective Removal of Zinc from Tower Blowdown by Ion Exchange," *Ind. Water Eng.* **8,** 33 (1971).

36. Gallo, B. G. and J. M. Culotta. "Save on Plating System," *Water Wastes Eng.* **9**, A18-A19 (1972).
37. Zievers, J. F. and C. J. Novotny. "Recovery of Mixed Rinse Water by Ion Exchange," *Plating* **58**, 482-485 (1971).
38. "Cost of Wastewater Treatment Processes," Robert A. Taft Research Center, Report No. TWRC-6, (Washington, D.C.: U.S. Dept. of the Interior, 1968).

INDEX

INDEX